Astronomy and Astrophysics Series Volume 7

The Virial Theorem in Stellar Astrophysics

Astronomy and Astrophysics Series
General Editor: A.G. Pacholczyk

Volume 1: T.L. Swihart
    Basic Physics of Stellar Atmospheres
Volume 2: T.L. Swihart
    Physics of Stellar Interiors
Volume 3: R.J. Weymann, T.L. Swihart, R.E. Williams, W.J. Cocke.
A.G. Pacholczyk, J.E. Felten
    Lecture Notes on Introductory Theoretical Astrophysics
Volume 4: E.R. Craine
    A Handbook of Quaisistellar and BL Lacertae Objects
    (Reference Works in Astronomy)
Volume 5: A.G. Pacholczyk
    A Handbook of Radio Sources. Part I, Strong Extragalactic
    Sources, 0-11 Hours. (Reference Works in Astronomy)
Volume 6: V.N. Zharkov, V.P. Trubitsyn
    Edited by W.B. Hubbard
    Physics of Planetary Interiors
Volume 7: G.W. Collins, II
    The Virial Theorem in Stellar Astrophysics
Volume 8: E.R. Craine
    A Sky Survey of Infared Optical Stellar Objects
    (Reference Works in Astronomy)
Volume 9: M. Heller and D.J. Raine
    The Relativity of Space-Time

Journals

The Astronomy Quarterly
Edited by E.R. Craine

The Virial Theorem

in Stellar Astrophysics

George W. Collins, II

The Ohio State University

Pachart Publishing House

Tucson

Copyright © 1978 by the Pachart Corporation

No part of this book may be reproduced by any mechanical, photograohic, or electronic process, or in the form of a phonographic recording, nor may it be stored in a retrieval system, transmitted, or otherwise copied for public or private use without written permission of the publisher.

Library of Congress Catalog Number: 78-059206
International Standard Book Number: 0-912918-13-6

Pachart Publishing House
Box 42963
Tucson, Arizona   85733

To the kindness, wisdom, humanity and memory of

       D. Nelson Limber
            and
       Uco Van Wijk

Table of Contents

Preface

Introduction
- 1. A brief historical review — 1
- 2. The nature of the theorem — 4
- 3. The scope and structure of the book — 6
- References — 7

Chapter I Development of the Virial Theorem
- 1. The basic equations of structure — 8
- 2. The classical derivation of the Virial Theorem — 11
- 3. Velocity dependent forces and the Virial Theorem — 15
- 4. Continuum-Field representation of The Virial Theorem — 16
- 5. The Ergodic Theorem and the Virial Theorem — 18
- 6. Summary — 23
- Notes to Chapter 1 — 25
- References — 26

Chapter II Contemporary Aspects of the Virial Theorem
- 1. The Tensor Virial Theorem — 27
- 2. Higher Order Virial Equations — 30
- 3. Special Relativity and the Virial Theorem — 34
- 4. General Relativity and the Virial Theorem — 37
- 5. Complications - Magnetic Fields, Internal Energy, and Rotation — 47
- 6. Summary — 54
- Notes to Chapter 2 — 57
- References — 60

Chapter III The Variational Form of the Virial Theorem
- 1. Variations, Perturbations, and their implications for the Virial Theorem — 61
- 2. Radial pulsations for self-gravitating systems — 63
- 3. The influence of magnetic and rotational energy upon a pulsating system — 68
- 4. Variational form of the surface terms — 78
- 5. The Virial Theorem and stability — 83
- 6. Summary — 95
- Notes to Chapter 3 — 97
- References — 102

Chapter IV Some Applications of the Virial Theorem
- 1. Pulsational stability of White Dwarfs — 103
- 2. The influence of rotation and magnetic fields on the White Dwarf Gravitational Instability — 111
- 3. Stability of Neutron Stars — 116
- 4. Additional topics and final thoughts — 120
- Notes to Chapter 4 — 127
- References — 128

Appendix Symbol Definitions and first usage — 129

Index — 134

Preface

As Fred Hoyle has observed, most readers assume a preface is written first and thus contains the authors hopes and aspirations. In reality most prefaces are written after the fact and contain the authors' views of his accomplishments. So it is in this case and I am forced to observe that my own perception of the subject has deepened and sharpened the considerable respect I have always had for the virial theorem. A corollary aspect of this expanded perspective is an awareness of how much remains to be done. Thus by no means can I claim to have prepared here a complete and exhaustive discussion of the virial theorem, rather this effort should be viewed as a guided introduction, punctuated by a few examples. I can only hope that the reader will proceed with the attitude that this constitutes not an end in itself, but an establishment of a point of view that is useful in comprehending some of the aspects of the universe.

A second traditional role of a preface is to provide a vehicle for acknowledging the help and assistance the author received in the preparation of his work. In addition to the customary accolades for proof reading which in this instance go to George Sonneborn and Dr.

John Faulkner, and manuscript preparation by Mrs. Delores Chambers, I feel happily compelled to heap praise upon the publisher.

It is not generally appreciated that there are only a few thousand astronomers in the United States and perhaps twice that number in the entire world. Only a small fraction of these could be expected to have an interest in such an apparently specialized subject. Thus the market for such a work compared to a similar effort in another domain of physical sciences such as Physics, Chemistry or Geology is miniscule. This situation has thereby forced virtually all contemporary thought in astrophysics into the various journals, which for economic reasons similar to those facing the would-be book publisher, find little room for contempletive or reflective thought. So it is a considerable surprise and great pleasure to find a publisher willing to put up with such problems and produce works of this type for the small but important audience that has need of them.

Lastly I would like to thank my family for trying to understand why anyone would write a book that won't make any money.

George W. Collins, II

The Ohio State University
November 15, 1977

Introduction

1. A Brief Historical Review

    Although most students of physics will recognize the name of the Virial Theorem, few can state it corrrecetly and even fewer appreciate its power. This is largely the result of its diverse development and somewhat obscure origin, for the virial theorem did not spring full blown in its present form but rather evolved from the studies of the kinetic theory of gases. One of the lasting achievements of 19th century physics was the development of a comprehensive theory of the behavior of confined gases which resulted in what is now known as thermodynamics and statistical mechanics. A brief, but impressive, account of this historical development can be found in "The Dynamical Theory of Gases" by Sir James Jeans[1] and in order to place the virial theorem in its proper prospective, it is worth recounting some of that history.
    Largely inspired by the work of Carnot on heat engines, R. J. E. Claussius began a long study of the mechanical nature of heat in 1851[2]. This study led him through twenty years to the formulation of what we can now see to be the earliest clear presentation of the

virial theorem. On June 13, 1870, Claussius delivered a lecture before the Association for Natural and Medical Sciences of the Lower Rhine "On a Mechanical Theorem Applicable to Heat."[3] In giving this lecture, Claussius stated the theorem as "The mean vis viva of the system is equal to its virial."[4]

In the 19th century, it was commonplace to assign a Latin name to any special characteristic of a system. Thus, as is known to all students of celestial mechanics the vis viva integral is in reality the total kinetic energy of the system. Claussius also turned to the Latin word virias (the plural of vis) meaning forces to obtain his name for the term involved in the second half of his theorem. This scalar quantity which he called the virial can be represented in terms of the forces acting on the system $\mathbf{F}_i$ as $1/2 \; \overline{\sum_i \mathbf{F}_i \cdot \mathbf{r}_i}$ and can be shown to be 1/2 the average potential energy of the system. So, in the more contemporary language of energy, Claussius would have stated that the average kinetic energy is equal to 1/2 the average potential energy. Although the characteristic of the system Claussius called the virial is no longer given much significance as a physical concept, the name has become attached to the theorem and its evolved forms.

Even though Claussius' lecture was translated and published in Great Britain in a scant six weeks, the power of the theorem was slow in being recognized. This lack of recognition prompted James Clerk Maxwell four years later to observe that "as in this country the importance of this theorem seems hardly to be appreciated, it may be as well to explain it a little more fully."[5] Maxwell's observation is still appropriate over a century later and indeed serves as the "raison d'etre" for this book.

After the turn of the century the applications of the theorem became more varied and widespread. Lord Rayleigh formulated a generalization of the theorem in 1903[6] in which one can see the beginnings of the tensor virial theorem revived by Parker[7] and later so extensively developed by Chandrasekhar during the 1960's.[8] Poincare used a form of the virial theorem in 1911[9] to investigate the stability of structures in different cosmological theories. During the 1940's Paul Ledoux developed a variational form of the virial theorem to obtain pulsational periods for stars and investigate their stability.[10] Chandrasekhar and Fermi extended the virial theorem in 1953 to include the presence of magnetic fields.[11]

At this point astute students of celestial mechanics will observe that the virial theorem can be obtained directly from Lagrange's Identity by simply averaging it over time and making a few statements concerning the stability of the system. Indeed, it is this derivation which is most often used to establish the virial theorem. Since Lagrange predates Claussius by a century, some comment is in order as to who has the better claim to the theorem.

In 1772 the Royal Academy of Sciences of Paris published J. L. Lagrange's "Essay on the Problem of Three Bodies."[12] In this essay he developed what can be interpreted as Lagrange's Identity for three bodies. Of course terms such as "moment of inertia", "potential" and "kinetic energy" do not appear but the basic mathematical formulation is present. It does appear that this remained a special case germain to the three-body problem until the winter of 1842-43 when Karl Jacobi generalized Lagrange's result to n-bodies. Jacobi's formulation closely parallels the present representation of Lagrange's identity including the relating of what will later be known as the virial of Claussius to the potential.[13] He continues on in the same chapter to develop the stability criterion for n-body systems which bears his name. It is indeed a very short step from this point to what is known as the Classical Virial Theorem. It is difficult to imagine that the contemporary Claussius was unaware of this work. However, there are some notable and important differences between the virial theorem of Claussius and that which can be deduced from Jacobi's formulation of Lagrange's Identity. These differences are amplified by considering the state of physics during the last half of the 19th century. The passion for unification which pervades 20th century physics was not extant in the time of Jacobi and Claussius. The study of heat and classical dynamics of gravitating systems were regarded as two very distinct disciplines. The formulation of statistical mechanics which now provides some measure of unity between the two had not been accomplished.

In addition the characterization of the properties of a gas in terms of its internal and kinetic energy was not yet developed. The very fact that Claussius required a new term, the virial, for the theorem makes it clear that its relationship to the internal energy of the gas was not clear. In addition, although he makes use of time averages in deriving the theory, it is clear from the development that he expected these averages to be interpreted as phase or ensemble averages. It is this last point which provides a major distinction between the virial theorem of Claussius and that obtainable from Lagrange's Identity. The point is subtle and often overlooked today, for only if the system is ergodic (in the sense of obeying the ergodic theorem) are phase and time averages the same. We will return to this point later in some detail. Thus it is fair to say that although the dynamical foundation for the virial theorem existed well before Claussius' pronouncement, by demonstrating its applicability to thermodynamics he made a new and fundamental contribution to physics.

2. The Nature of the Theorem

By now the reader may have gotten some feeling for the wide ranging applicability of the virial theorem. Not only is it applicable to dynamical and thermodynamical systems, but we shall see that it can also be formulated to deal with relativistic (in the sense of special relativity) systems, systems with velocity dependent forces, viscous systems, systems exhibiting macroscopic motions such as rotation, systems with magnetic fields and even systems which require general relativity for their description. Since the theorem represents a basic structural relationship which the system must obey, applying the Calculus of Variations to the theorem can be expected to provide information regarding its dynamical behavior and the way in which the presence of additional phenomena (e.g., rotation, magnetic fields, etc.) affect that behavior.

Let us then prepare to examine why this theorem can provide information concerning systems whose complete analysis may defy description. Within the framework of classical mechanics, most of the systems I mentioned above can be described by solving the force equations representing the system. These equations can usually be obtained from the beautiful formalisms of Lagrange or Hamilton or from the Boltzmann transport equation. Unfortunately, those equations will, in general, be non-linear, second-order, vector differential equations which only in special cases exhibit closed form solutions. Although additional cases may be solved numerically, insight into the behavior of systems in general is very difficult to obtain in this manner. However, the virial theorem generally deals in scalar quantities and usually is applied on a global scale. It is indeed this reduction in complexity from a vector description to a scalar one which enables us to solve the resulting equations. This reduction results in a concomitant loss of information and we cannot expect to obtain as complete a description of a physical system as would be possible from the solution of the force equations.

There are two ways of looking at the reason for this inability to ascertain the complete physical structure of a system from energy considerations alone. First, the number of separate scalar equations one has at his disposal are fewer in the energy approach than in the force approach. That is, the energy considerations yield equations involving only energies or 'energy-like' scalars while the force equations being vector equations, yield at least three separate 'component' equations which in turn will behave as coupled scalar equations. One might sum up this argument by simply saying that there is more information contained in a vector than in a scalar.

The second method of looking at the problem is to note that energies are normally first integrals of forces. Thus the equations we shall be primarily concerned with are related to the first integral of the defining differential force equations. Now when one integrates a function, he loses 'information' about the function. That is, he lumps up the detailed structure of the function over a discrete range into a single quantity known as integral of the function, and in doing so loses any knowledge of that detailed structure. Therefore, since the process of integration results in a loss of information, we cannot expect the energy equation (representing the first integral of the force equation) to yield as complete a picture of the system as does the solution of the force equations themselves.

However, this loss of detailed structure is somewhat compensated for, firstly by being able to solve the resulting equations due to their greater simplicity, and secondly, by being able to consider more difficult problems whose formulation in detail is at present beyond the scope of contemporary physics.

3. The Scope and Structure of the Book

Any introduction to a book would be incomplete if it failed to delimit its scope. Initially one might wonder at such an extensive discussion of a single theorem. In reality it is not possible to cover in a single text all of the diverse applications and implications of this theorem. All areas of physical science in which the concepts of force and energy are important are touched by the virial theorem. Even within the more restricted study of astronomy, the virial theorem finds applications in the dust and gas of interstellar space and cosmological considerations of the universe as a whole. Restriction of this investigation to stars and stellar systems would admit discussions concerning the stability of clusters, galaxies and clusters of galaxies which could in themselves fill many separate volumes. Thus, we shall primarily concern ourselves with the application of the virial theorem to the astrophysics of stars and star-like objects. Indeed, since research into these objects is still an open and aggressively pursued subject, I shall not even be able to guarantee that this treatment is complete and comprehensive. Since, as I have already noted, the virial theorem does not by its very nature provide a complete description of a physical system but rather extensive insight into its behavior, let me hope that this same spirit of incisive investigation will pervade the rest of this work.

With regard to the organization and structure of what follows, let me emphasize that this is a book for students - young and old. To that end, I have endeavored to avoid such phrases as - "it can easily be shown that . . . .", or others designed to extol the intellect of the author at the expense of the reader. Thus, in an attempt to clarify the development I have included most of the algebraic steps of the development. The active professional or well prepared student may skip many of these steps without losing content or continuity. The skeptic will wish to read them all. However, in order not to burden the casual reader, the more tedious algebra has been relegated to notes which may be found at the end of each chapter. Each chapter of the book has been subdivided into sections (as has the introduction), which represent a particular logically cohesive unit. At the end of each chapter, I have chosen to provide a brief summary of what I feel constitutes the major thread of that chapter. A comfortable rapport with the content of these summaries may encourage the reader in the belief that he is understanding what the author intended.

REFERENCES

1 Jeans, J. H. 1925 Dynamical Theory of Gases, Cambridge University Press, London, p. 11.
2 Claussius, R. J. E. 1851. Phil. Mag. S. 4 Vol. 2, p. 1-21, 102-119.
  Claussius, R. J. E. 1856. Phil. Mag. S. 4, Vol. 12, p. 81.
  Claussius, R. J. E. 1862. Phil. Mag. S. 4, Vol. 24, p. 81-97, 201-213.
3 Claussius, R. J. E. 1870. Phil. Mag. S. 4, Vol. 40, p. 122.
4 Claussius, R. J. E. 1870. Phil. Mag. S. 4, Vol. 40, p. 124.
5 Maxwell, J. C. 1874. Scientific Papers. Vol. 2, p. 410, Dover Publications, Inc., N. Y.
6 Rayleigh, L. 1903. Scientific Papers. Vol. 4, p. 491, Cambridge, England.
7 Parker, E. N. 1954. Phys. Rev. 96, p. 1686-9.
8 Chandrasekhar, S. 1960.
9 Poincare, H. 1911. Lectures on Cosmological Theories, Hermann, Paris.
10 Ledoux, P. 1945, Ap. J. 102, p. 134-153.
11 Chandrasekhar, S. and Fermi, E. 1953, Ap. J. 118, p.116.
12 Lagrange, J. L. 1873. Oeuvres de Lagrange ed: M. J.-A. Serret Gauthier-Villars, Paris, p. 240, 241.
13 Jacobi, C. G. J. 1889. Varlesungen uber Dynamik ed: A. Clebsch G. Reimer, Berlin, p. 18-22.

1. Development of the Virial Theorem

1. The Basic Equations of Structure

Before turning to the derivation of the virial theorem, it is appropriate to review the origin of the fundamental structural equations of stellar astrophysics. This not only provides insight into the basic conservation laws implicitly assumed in the description of physical systems, but by their generality and completeness graphically illustrates the complexity of complete description that we seek to circumvent. Since lengthy and excellent texts already exist on this subject, our review will of necessity be a sketch. Any description of a physical system begins either implicitly or explicitly from certain general conservation principles. Such a system is considered to be a collection of particles, each endowed with a spatial location and momentum which move under the influence of known forces. If one regards the characteristics of spatial position and momentum as being highly independent, then one can construct a multi-dimensional space through which the particles will trace out unique paths describing this history.

This is essentially a statement of determinism, and in classical terms is formulated in a six-dimensional space called phase-space consisting of three spatial dimensions and three linearly independent momentum dimensions. If one considers an infinitesimal volume of this space, he may formulate a very general conservation law which simply says that the divergence of the flow of particles in that volume is equal to the number created or destroyed within that volume. The mathematical formulation of this concept is usually called the Boltzmann transport equation and takes this form:

$$\frac{\partial \psi}{\partial t} + \sum_{i=1}^{3} \dot{x}_i \frac{\partial \psi}{\partial x_i} + \sum_{i=1}^{3} \dot{p}_i \frac{\partial \psi}{\partial p_i} = S \qquad 1.1.1$$

or in vector notation

$$\frac{\partial \psi}{\partial t} + \mathbf{v} \cdot \nabla \psi + \mathbf{f} \cdot \nabla_p \psi = S$$

where ψ is the density of points in phase space, $\mathbf{f}$ is the sum of the forces acting on the particles and S is the 'creation rate' of particles within the volume. The homogeneous form of this equation is often called the Louisville Theorem and would be discussed in detail in any good book on classical mechanics.

A determination of ψ as a function of the coordinates and time constitutes a complete description of the system. However, rarely is an attempt made to solve equation 1.1.1 but rather simplifications are made from which come the basic equations of stellar structure. This is generally done by taking 'moments' of the equations with respect to the various coordinates. For example, noting that the integral of ψ over all velocity space yields the matter density ρ and that no particles can exist with unbounded momentum, averaging equation 1.1.1 over all velocity space yields

$$\frac{\partial \rho}{\partial t} + \nabla \cdot (\mathbf{u} \rho) = \overline{S} \qquad 1.1.2$$

where $\mathbf{u}$ is the average stream velocity of the particles defined by

$$\mathbf{u} = \frac{1}{\rho} \int \psi \mathbf{v} \, dv \qquad 1.1.3$$

For systems where mass is neither created or destroyed $\bar{S} = 0$ and equation 1.1.2 is just a statement of the conservation of mass.

If one multiplies equation 1.1.1 by the particle velocities and averages again over all velocity space he will obtain after a great deal of algebra the Euler-Lagrange equations of hydrodynamic flow

$$\frac{\partial \mathbf{u}}{\partial t} + (\mathbf{u} \cdot \nabla) \mathbf{u} = -\nabla \Psi - \frac{1}{\rho} \nabla \cdot \mathbf{P} - \frac{1}{\rho} \int \mathbf{s}(\mathbf{v} - \mathbf{u}) dv \qquad 1.1.4$$

Here the forces $\mathbf{f}$ have been assumed to be derivable from a potential Ψ.

The symbol $\mathbf{P}$ is known as the pressure tensor and has the form

$$\mathbf{P} = \int \psi (\mathbf{v} - \mathbf{u})(\mathbf{v} - \mathbf{u}) dv \qquad 1.1.5$$

These rather formidable equations simplify considerably in the case where many collisions randomize the particle motion with respect to the mean stream velocity $\mathbf{u}$. Under these conditions the last term on the right of 1.1.4 vanishes and the pressure tensor becomes diagonal with each element equal. Its divergence then becomes the gradient of the familiar scalar known as the gas pressure. If we further consider only systems exhibiting no stream motion we arrive at the familiar equation of hydrostatic equilibrium

$$\nabla P = -\rho \nabla \Psi \qquad 1.1.6$$

Multiplying 1.1.1 by $\mathbf{v}$ and averaging over $\mathbf{v}$, has essentially turned the Boltzmann transport equation into an equation expressing the conservation of momentum. Equation 1.1.6 along with Poisson's equation for the sources of the potential

$$\nabla^2 \Psi = -4\pi G \rho \qquad 1.1.7$$

constitute a complete statement of the conservation of momentum.

Multiplying 1.1.1 by $\mathbf{v} \cdot \mathbf{v}$ or v^2 and averaging over all velocity space will produce an equation which represents the conservation of energy which when combined with the ideal gas law is

$$\rho \frac{dE}{dt} + p\nabla \cdot \mathbf{v} = \rho\varepsilon + \chi - \nabla \cdot \mathbf{F} \qquad 1.1.8$$

where $\mathbf{F}$ is the radiant flux, ε the total rate of energy generation and χ is the energy generated by viscous motions. If one has a machine wherein no mass motions exist and all energy flows by radiation, we have

$$\nabla \cdot \mathbf{F} = \rho\varepsilon \qquad 1.1.9$$

For static configurations exhibiting spherical symmetry these conservative laws take their most familiar form:

Conservation of mass $\qquad \dfrac{dm(r)}{dr} = 4\pi r^2 \rho$

Conservation of momentum $\qquad \dfrac{dP(r)}{dr} = -\dfrac{Gm(r)\rho}{r^2}$

Conservation of energy $\qquad \dfrac{dL(r)}{dr} = 4\pi r^2 \rho\varepsilon \qquad 1.1.10$

$\qquad\qquad\qquad\qquad\qquad L(r) = 4\pi r^2 F$

2. The Classical Derivation of the Virial Theorem

The virial theorem is often stated in slightly different forms having slightly different interpretations. In general, we shall repeat the version given by Claussius and express the virial theorem as a relation between the average value of the kinetic and potential energies of a system in a steady state or a quasi-steady state. Since the understanding of any theorem is related to its origins, we shall spend some time deriving the virial theorem from first principles. Many derivations of varying degree of completeness exist in the literature. Most texts on stellar or classical dynamics (e.g. Kurth)[1] derive the theorem from the Lagrange identity. Landau and Lifshitz[2] give an eloquent derivation appropriate for the electromagnetic field which we shall consider in more detail in the next section. Chandrasekhar[3] follows closely the approach of Claussius while Goldstein[4] gives a very readable vector derivation firmly rooted in the original approach and it is basically this form we shall develop first.

Consider a general system of mass points m_i with position vectors $\mathbf{r}_i$ which are subjected to applied forces (including any forces of constraint) $\mathbf{f}_i$. The Newtonian equations of motions for the system are then

$$\mathbf{p}_i = \frac{d}{dt}(m_i \mathbf{v}_i) = \mathbf{f}_i \qquad 1.2.1$$

Now define
$$G = \sum_i \dot{\mathbf{p}}_i \cdot \mathbf{r}_i = \sum_i m_i \frac{d\mathbf{r}_i}{dt} \cdot \mathbf{r}_i$$

$$= \frac{1}{2} \sum_i m_i \frac{d}{dt}(\mathbf{r}_i \cdot \mathbf{r}_i)$$

or
$$G = \frac{1}{2} \frac{d}{dt}\left[\sum_i m_i r_i^2\right]. \qquad 1.2.2$$

The term in the square brackets is the moment of inertia (by definition) about a point and that point is the origin of the coordinate system for the position vectors $\mathbf{r}_i$. Thus, we have

$$G = 1/2 \frac{dI}{dt} \qquad 1.2.3$$

where I is the moment of inertia about the origin of the coordinate system.

Now consider

$$\frac{dG}{dt} = \sum_i \dot{\mathbf{r}}_i \cdot \mathbf{p}_i + \sum_i \dot{\mathbf{p}}_i \cdot \mathbf{r}_i \qquad 1.2.4$$

but
$$\sum_i \dot{\mathbf{r}}_i \cdot \mathbf{p}_i = \sum_i m_i \dot{\mathbf{r}}_i \cdot \dot{\mathbf{r}}_i =$$

$$\sum_i m_i v_i^2 = 2T \qquad 1.2.5$$

where T is the total kinetic energy of the system with respect to the origin of the coordinate system. However, since $\dot{\mathbf{p}}_i$ is really the applied force acting on the system (see equation 1.2.1), we may rewrite (1.2.4) as follows:

$$\frac{dG}{dt} = 2T + \sum_i \mathbf{f}_i \cdot \mathbf{r}_i \quad . \qquad 1.2.6$$

The last term on the right is known as the Virial of Claussius.

Now consider the Virial of Claussius. Let us assume that the forces $\mathbf{f}_i$ obey a power law with respect to distance and are derivable from a potential. The total force on the ith particle may be determined by summing all the forces acting on that particle. Thus

$$\mathbf{f}_i = \sum_{j \neq i} \mathbf{F}_{ij} \quad , \qquad 1.2.7$$

where $\mathbf{F}_{ij}$ is the force between the ith and jth particle. Now, if the forces obey a power law and are derivable from a potential then,

$$\mathbf{F}_{ij} = \nabla_i m_i \Phi(r_{ij}) = -\nabla_i a_{ij} r_{ij}^n \qquad 1.2.8$$

The subscript on the ∇ operator implies that the gradient is to be taken in a coordinate system having the ith particle at the origin. Carrying out the operation implied by (1.2.8), we have

$$\mathbf{F}_{ij} = -n\, a_{ij}\, r_{ij}^{(n-2)} (\mathbf{r}_i - \mathbf{r}_j) \qquad 1.2.9$$

Now since the force acting on the ith particle due to the jth particle may be paired off with a force exactly equal and oppositely directed, acting on the jth particle due to the ith particle, we can rewrite 1.2.7 as follows:

$$\mathbf{f}_i = \sum_{j=1} \mathbf{F}_{ij} = \sum_{j>i} \mathbf{F}_{ij} + \mathbf{F}_{ji} \qquad 1.2.10$$

Substituting (1.2.10) into the definition of the virial of Claussius, we have

$$\sum_i \mathbf{f}_i \cdot \mathbf{r}_i = \sum_i \sum_{j>i} \mathbf{F}_{ij} \cdot \mathbf{r}_i + \mathbf{F}_{ij} \cdot \mathbf{r}_j \qquad 1.2.11$$

It is important here to notice that the position vector $\mathbf{r}_i$, which is 'dotted' into the force vector, bears the same subscript as the first subscript on the force vector.

That is, the position vector is the vector from the origin of the coordinate system to the particle being action upon. Substitution of 1.2.8 into 1.2.11 and then into 1.2.6 yields:

$$\frac{dG}{dt} = 2T - n\mathcal{U} \quad . \qquad 1.2.12$$

where $\mathcal{U}$ is the total potential energy.[1.1]
For the gravitational potential n = -1, and we arrive at a statement of what is known as Lagranges' Identity

$$\frac{dG}{dt} = \frac{1}{2}\frac{d^2 I}{dt^2} = 2T + \Omega \quad . \qquad 1.2.13$$

To arrive at the usual statement of the virial theorem we must average over an interval of time (T_o). It is in this sense that the virial theorem is sometimes referred to as a statistical theorem. Therefore, integrating equation (1.2.12), we have

$$\frac{1}{T_o}\int_0^{T_o}\frac{dG}{dt}dt = \frac{2}{T_o}\int_0^{T_o}T(t)dt - \frac{n}{T_o}\int_0^{T_o}\mathcal{U}(t)dt \qquad 1.2.14$$

and, using the definition of average value we obtain:

$$\frac{1}{T_o}[G(T_o) - G(o)] = 2\overline{T} - n\overline{\mathcal{U}} \quad . \qquad 1.2.15$$

If the motion of the system over a time T_o is periodic, then the left-hand side of (1.2.15) will vanish. Indeed, if the motion of the system is bounded (i.e., $G(t) < \infty$), then we may make the left hand side of (1.2.15) as small as we wish by averaging over a longer time. Thus, if a system is in a steady state the moment of inertia (I) is constant and for systems governed by gravity

$$2\overline{T} + \overline{\Omega} = 0 \quad . \qquad 1.2.16$$

It should be noted that this formulation of the virial theorem involves time averages of indeterminate length. If one is to use the virial theorem to determine whether a system is in accelerative expansion or contraction, then he must be very careful about how he obtains the average value of the kinetic and potential energies.

3. Velocity Dependent Forces and the Virial Theorem

There is an additional feature of the virial theorem as stated in (1.2.16) that should be mentioned. If the forces acting on the system include velocity dependent forces, the result of the virial theorem is unchanged. In order to demonstrate this, consider the same system of mass points m_i subjected to forces $\mathbf{f}_i$ which may be divided into velocity dependent ($\overline{\mathbf{w}}_i$) and velocity independent forces ($\mathbf{z}_i$). The equations of motion may be written as:

$$\dot{\mathbf{p}}_i = \mathbf{f}_i = \overline{\mathbf{w}}_i + \mathbf{z}_i \qquad 1.3.1$$

Substituting into (1.2.6), we have

$$\frac{dG}{dt} - \sum_i \overline{\mathbf{w}}_i \cdot \mathbf{r}_i = 2T + \sum_i \mathbf{z}_i \cdot \mathbf{r}_i \qquad 1.3.2$$

Remembering that the velocity dependent forces may be rewritten as

$$\overline{\mathbf{w}}_i = \alpha_i \mathbf{v}_i = \alpha_i \frac{d\mathbf{r}_i}{dt} \qquad 1.3.3$$

We may again average over time as in equation (1.2.12). Thus

$$\frac{1}{T_o} \int_o^{T_o} \frac{dG}{dt} dt - \frac{1}{T_o} \int_o^{T_o} \sum_i \alpha_i \frac{d\mathbf{r}_i}{dt} \cdot \mathbf{r}_i \, dt$$

$$= 2\overline{T} - n\overline{\mathcal{U}} \qquad 1.3.4$$

where $\overline{\mathcal{U}}$ is the average value of the potential energy for the "non-frictional" forces. Carrying out the integration on the left hand side we have

$$\frac{1}{T_o}[G(T_o) - G(o)] + \frac{1}{2T_o} \sum_i \alpha_i [r_i^2(T_o) - r_i^2(o)]$$

$$= 2\overline{T} - n \qquad 1.3.5$$

Thus, if the motion is periodic, both terms on the left hand side of equation (1.3.5) will vanish in a time T_o equal to the period of the system. Indeed both terms can be made as small as required providing the "frictional" forces $\overline{\mathbf{w}}_i$ do not cause the system to cease to be in motion over the time for which the averaging is done. This apparently academic aside has the significant result that we need not worry about any Lorentz forces or viscosity forces which may be present in our subsequent discussion in which we shall invoke the virial theorem.

4. Continuum-Field Representation of the Virial Theorem

Although nearly all derivations of the virial theorem consider collections of mass-points acting under forces derivable from a potential, it is useful to look at this formalism as it applies to a continuum density field of matter. This is particularly appropriate when one considers applications to stellar structure where a continuum representation of the material is always used.

In the interests of preserving some rigor let us pass from equation 1.2.1 to its analogous representation in the continuum. Let the mass m_i be obtained by multiplying the density $\rho(r)$ by an infinitismal volume ΔV so that 1.2.1 becomes

$$\mathbf{f}_i = \frac{d}{dt}(\rho \mathbf{v} \Delta V) = \mathbf{v} \Delta V \frac{d\rho}{dt} + \rho \Delta V \frac{d\mathbf{v}}{dt} + \rho \mathbf{v} \frac{d(\Delta V)}{dt} \qquad 1.4.1$$

Conservation of mass requires that

$$\frac{dm_i}{dt} = \frac{d}{dt}(\rho \Delta V) = \Delta V \frac{d\rho}{dt} + \rho \frac{d(\Delta V)}{dt} = 0 \qquad 1.4.2$$

Multiplying this expression by $\mathbf{v}$ we see that the first and last terms on the right hand side of 1.4.1 are of equal magnitude and opposite sign. Thus, if we define a "force-density", $\mathbf{f}$ so that $\mathbf{f} \Delta V = \mathbf{f}_i$, we can pass to this continuum representation of 1.2.1

$$\mathbf{f}(\mathbf{r}) = \rho(\mathbf{r}) \frac{d}{dt}\left[\mathbf{v}(\mathbf{r})\right] = \dot{\mathbf{p}}(\mathbf{r}) \qquad 1.4.3$$

where $\mathbf{p}(r)$ by analogy to 1.2.1 is just the local momentum density.

We can now define G in terms of the continuum variables so that

$$G = \int_V \mathbf{p} \cdot \mathbf{r} \, dV = \int_V \rho \frac{d\mathbf{r}}{dt} \cdot \mathbf{r} \, dV$$

$$= \frac{1}{2} \int_V \rho \frac{d}{dt}(\mathbf{r} \cdot \mathbf{r}) dV = \frac{1}{2} \int_V \rho \frac{dr^2}{dt} dV \qquad 1.4.4$$

$$G = \frac{1}{2} \int_V \frac{d}{dt}(\rho r^2) dV - \frac{1}{2} \int_V r^2 \frac{d\rho}{dt} dV \qquad 1.4.5$$

Once again, one uses conservation of mass requiring that the mass within any sub volume V' is constant with time so that $\frac{dm(V')}{dt} = 0$ with that sub volume V' defined such that

$$\frac{d}{dt}\int_{V'} \rho dV = \int_{V'} \frac{d\rho}{dt} dV = 0 \qquad 1.4.6$$

Thus, the second integral in 1.4.5 after integration by parts is zero. If we take the original volume V to be large enough so as to always include all the mass of the object, we may write 1.4.5 as

$$G = \frac{1}{2} \frac{d}{dt} \int_V (\rho r^2) dV = \frac{1}{2} \frac{dI}{dt} \quad . \qquad 1.4.7$$

With these same constraints on V we may differentiate 1.4.4 with respect to time and obtain

$$\frac{dG}{dt} = \int_V [\boldsymbol{p} \cdot \frac{d\boldsymbol{r}}{dt} + \boldsymbol{r} \cdot \frac{d\boldsymbol{p}}{dt}] dV$$
$$= \int_V (\rho v^2 + \boldsymbol{r} \cdot \boldsymbol{f}) dV \quad . \qquad 1.4.8$$

The first term under the integral is just kinetic energy density and hence its volume integral is just the total kinetic energy of the configuration and

$$\frac{1}{2} \frac{d^2 I}{dt^2} = 2T + \int_V \boldsymbol{f} \cdot \boldsymbol{r} \, dV \quad . \qquad 1.4.9$$

Considerable care must be taken in evaluating the second term in equation 1.4.9 which is basically the virial of Claussius. In the previous derivation we went to some length (i.e., equation 1.2.10) to avoid "double counting" the forces by noting that the force between any two particles A and B can be viewed as a force at A due to B, or a force at B resulting from A. The contributions to the virial, however, are not equal as they involve a 'dot' product with the position vector. Thus, we explicitly paired the forces and arranged the sum so pairs of particles were only counted once. Similar problems confront us within continuum derivation. Thus, each force at a field point $\boldsymbol{f}(\boldsymbol{r})$ will have an equal and opposite counterpart at the source points $\boldsymbol{r}'$.

After some algebra direct substitution of the potential gradient into the definition of the Virial of Claussius yields[1,2]

$$\int_V \mathbf{f} \cdot \mathbf{r} \, dV = -\frac{n}{2} \int_V \int_{V'} \rho(\mathbf{r})\rho(\mathbf{r}')(\mathbf{r}-\mathbf{r}')(\mathbf{r}-\mathbf{r}')(|\mathbf{r}-\mathbf{r}'|)^{n-2} dV dV'$$

$$= -n \left[\frac{1}{2} \int_V \int_{V'} \rho(\mathbf{r}) \rho(\mathbf{r}') (|\mathbf{r}-\mathbf{r}'|)^n dV' dV \right] \qquad 1.4.10$$

Since $V = V'$, the integrals are fully symmetric with respect to interchanging primed with non-primed variables. In addition the double integral represents the potential energy of $\rho(\mathbf{r})$ with respect to $\rho(\mathbf{r}')$, and $\rho(\mathbf{r}')$ with respect to $\rho(\mathbf{r})$; it is just twice the total potential energy. Thus, we find that the virial has the same form as equation 1.2.12.

$$\int_V \mathbf{f} \cdot \mathbf{r} \, dV = -n\, \mathcal{U} \qquad 1.4.11$$

Substitution of this form into 1.4.9 and taking $n = -1$ yields the same expression for Lagrange's identity as was obtained in 1.2.13, namely,

$$\frac{1}{2} \frac{d^2 I}{dt^2} = 2T + \Omega \qquad 1.4.12$$

Thus Lagrange's identity, the virial theorem and indeed the remainder of the earlier arguments, are valid for the continuum density distributions as we might have guessed.

Throughout this discussion it was tacitly assumed that the forces involved represented "gravitational" forces insofar as the force was $-\rho \nabla \Phi$. Clearly, if the force depended on some other property of the matter (e.g., the charge density, $\varepsilon(r)$ the evaluation of $\int_V \mathbf{f} \cdot \mathbf{r} \, dV$ would go as before with the result that the virial would again be $-n\,\mathcal{U}$ where $\mathcal{U}$ is the total potential energy of the configuration.

5. The Ergodic Theorem and the Virial Theorem

Thus far, with the exception of a brief discussion in Section 2, we have developed Lagrange's identity in a variety of ways, but have not rigorously taken that final step to produce the virial theorem. This last step in-

volves averaging over time and it is in this form that the
theorem finds its widest application. However, in astrophysics few, if any, investigators, live long enough to
perform the time-averages for which the theorem calls.
Thus, one more step is needed. It is this step which
occasionally leads to difficulty and erroneous results.
In order to replace the time averages with something observable, it is necessary to invoke the ergodic theorem.

The Ergodic Theorem is one of those fundamental
physical concepts like the Principle of Causality which
are so "obvious" as to appear axiomatic. Thus they are
rarely discussed in the physics literature. However, to
say that the ergodic theorem is obvious is to belittle an
entire area of mathematics known as ergodic theory which
uses the language of measure theory. This language alone
is enough to hide it forever from the eye of the average
physical scientist. Since this theorem is central to obtain what is commonly called the virial theorem, it is
appropriate that we spend a little time on its meaning.

As noted in the introduction, the distinction between an ensemble average and an average of macroscopic
system parameters over time was not clear at the time of
the formulation of the virial theorem. However, not too
long after, Ludwig Boltzmann[6] formulated an hypothesis
which suggested the criterion under which ensemble and
phase averages would be the same. Maxwell later stated
it this way: "The only assumption which is necessary for
a direct proof is that the system if left to itself in its
actual state of motion will, sooner or later, pass through
every phase which is consistent with the equation of energy".[7] Essentially this constitutes what is most commonly meant by the ergodic theorem. Namely, if a dynamic
system passes through every point in phase space then the
time average of any macroscopic system parameter, say Q,
is given by

$$\langle Q \rangle_t = \lim_{T \to \infty} \frac{1}{T} \int_{t_o}^{t_o+T} Q(t) dt = \langle Q \rangle_s \qquad 1.5.1$$

where $\langle Q \rangle_s$ is some sort of instantaneous statistical average
of Q over the entire system.

The importance of this concept for statistical mechanics is clear. Theoretical considerations predict $\langle Q \rangle_s$
whereas experiment provides something which might be construed to approximately $\langle Q \rangle_t$. No matter how rapid the

measurement of something like the pressure or temperature of the gas, it requires a time which is long compared to characteristic times for the system. The founders of statistical mechanics, such as Boltzmann, Maxwell and Gibbs, realized that such a statement as 1.5.1 was necessary to enable the comparison of theory with experiment and thus a great deal of effort was expended to show or at least define the conditions under which dynamical systems were ergodic (i.e., would pass through every point in phase space).

Indeed, as stated the ergodic theorem is false as was shown independently in 1913 by Rosenthal[8] and Plancherel[9]. A more modern version of this can be seen easily by noting that no system trajectory in phase space may cross itself. Thus, such a curve may have no multiple points. This is effectively a statement of system boundary conditions uniquely determining the system's past and future. It is the **essence** of the Louisville theorem of classical mechanics. Such a curve is topologically known as a Jordan curve and it is a well known topological theorem that a Jordan curve cannot pass through all points of a multi-dimensional space. In the language of measure theory, a multi-dimensional space filling curve would have a measure equal to the space whereas a Jordan curve being one-dimensional would have measure zero.

Thus, the ergodic hypothesis became modified as the quasi-ergodic hypothesis. This modification essentially states that although a single phase trajectory cannot pass through every point in phase space, it may come arbitrarily close to any given point in a finite time. Already one can sense confusion of terminology beginning to mount. Ogorodnikov[10] uses the term quasi-ergodic to apply to systems covered by the Lewis theorem which we shall mention later. At this point in time the mathematical interest in ergodic theory began to rise rapidly and over the next several years attracted some of the most famous mathematical minds of the 20th century. Farquhar[11] points out that several noted physicists stated without justification that all physical systems were quasi-ergodic. The stakes were high and were getting higher with the development of statistical mechanics and the emergence of quantum mechanisms as powerful physical disciplines. The identity of phase and time averages became crucial to the comparison of theory with observation.

Mathematicians largely took over the field developing the formidable literature currently known as ergodic theory; and they became more concerned with showing the existance of the averages than with their equality with phase averages. Physicists, impatient with mathematicians for being unable to prove what appears 'reasonable' and also what is necessary began to require the identity of phase and time averages as being axiomatic. This is a position not without precedent and a certain pragmatic justification of expediency. Some essentially adopted the attitude that since thermodynamics "works", phase and times averages must be equal. However, as Farquhar observed - "such a pragmatic view reduces statistical mechanics to an ad hoc technique unrelated to the rest of physical theory."[12]

Over the last half century, there have been many attempts to prove the quasi-ergodic hypothesis - the most notable of which are Birkhoff's theorem[13] and the generalization of a corallory known as Lewis' theorem.[14] These theorems show the existence of time averages and their equivalence to phase averages under quite general conditions. The tendency in recent years has been to bypass phase space filling properties of a dynamical system and go directly to the identification of the equality of phase and time averages. The most recent attempt due to Siniai[15], as recounted by Arnold and Avez[16] proves that the Boltzmann-Gibbs conjecture is correct. That is, a "gas" made up of perfectly elastic spheres confined by a container with perfectly reflecting walls is ergodic in the sense that phase and time averages are equal.

At this point the reader is probably wondering what all this has to do with the virial theorem. Specifically, the virial theorem is obtained by taking the time average of Lagrange's identity. Thus

$$\lim_{T \to \infty} \frac{1}{2} \int_{t_o}^{t_o+T} \left(\frac{d^2 I}{dt^2}\right) dt = <2T>_t - <\mathcal{U}>_t \qquad 1.5.2$$

and for systems which are stable the left hand side is zero. The first problem arises with the fact that the time average is over infinite time and thus operationally difficult to carry out[1,3] Farquhar[17] points out that the time interval must at least be long compared to the relaxation time for the system and in the event that the system crossing time is longer than the relaxation time, the integration in 1.5.2 must exceed that time if any statistical validity is to be maintained in the analysis of the system.

It is clear that for stars and star-like objects these conditions are met. However, in stellar dynamics and the analysis of stellar systems they generally are not. Indeed, in this case, the astronomer is in the envious position of being in the reverse position from the thermodynamicists. For all intents and purposes he can perform an 'instantaneous' ensemble average which he wishes to equate to a 'theoretically determined' time average. This interpretation will only be correct if the system is ergodic in the sense of satisfying the 'quasi-ergodic hypothesis'. Pragmatically if the system exhibits a large number of degrees of freedom then persuasive arguments can be made that the equating of time and phase averages is justified. However, if isolating integrals of the motion exist for the system, then it is not justified, as these integrals remove large regions of phase space from the allowable space of the system trajectory. Lewis' theorem allows for ergodicity in a sub-space but then the phase averages must be calculated differently and this correspondence to the observed ensemble average is not clear. Thus the application of the virial theorem to a system with only a few members and hence a few degrees of freedom is invalid unless care is taken to interpret the observed ensemble averages in light of phase averages altered by the isolating integrals of the motion. Furthermore, one should be most circumspect about applying the virial theorem to large systems like the galaxy which appear to exhibit quasi-isolating integrals of the motion. That is, integrals which appear to restrict the system motion in phase space over several relaxation times. However, for stars and star-like objects exhibiting 10^{50} or more particles undergoing rapid collisions and having short relaxation times, these concerns do not apply and we may confidently interchange time and phase averages as they appear in the virial theorem. At least we may do it with the same confidence of the thermodynamicist.

For those who feel that the ergodic theorem is still "much ado about nothing", it is worth observing that by attempting to provide a rational development between dynamics and thermodynamics ergodic theory must address itself to the problems of irreversible processes. Since classical dynamics is fully reversible and thermodynamics includes processes which are not, the nature of irreversibility must be connected in some sense to that of ergodicity and thus to the very nature of time itself. Thus, anyone truly interested in the foundations of physics cannot dismiss ergodic theory as mere mathematical 'nit-picking'.

6. Summary

In this chapter, I have tried to lay the groundwork for the classical virial theorem by first demonstrating its utility, then deriving it in several ways and lastly, examining an important premise of its application. An underlying thread of continuity can be seen in all that follows, from the Boltzmann transport equation. It is a theme that will return again and again throughout this book. In section 1, we sketched how the Boltzmann transport equation yields a set of conservation laws which in turn supply the basic structure equations for stars. This sketch was far from exhaustive and intended primarily to show the informational complexity of this form of derivation. Being suitably impressed with this complexity, the reader should be in an agreeable frame of mind to consider alternative approaches to solving the vector differential equations of structure in order to glean insight into the behavior of the system. The next two sections were concerned with a highly classical derivation of the virial theorem with section 2 being basically the derivation as it might have been presented a century ago. Section 3 merely updated this presentation so that the formalism may be used within the context of more contemporary field theory. The only 'tricky' part of these derivations involves the 'pairing' of forces. The reader should make every effort to understand or conceptualize how this occurs in order to understand the meaning of the virial itself. The assumption that the forces are derivable from a potential which is described by a power law of the distance alone, dates back at least to Jacobi and is often described as a homogeneous function of the distance.

In the last section, I attempted to provide some insight into the meaning of a very important theorem generally known as the ergodic theorem. Its importance for the application of the virial theorem cannot be too strongly emphasized. Although almost all systems of interest in stellar astrophysics can truly be regarded as ergodic, many systems in stellar dynamics cannot. If they are not, one cannot replace averages over time by averages over phase or the ensemble of particles without further justification.

Notes to Chapter 1

1.1 Since $a_{ij} = a_{ji}$ for all known physical forces, we may substitute (1.2.9) in (1.2.11) as follows:

$$\sum_i \mathbf{f}_i \cdot \mathbf{r}_i = -\sum_i \sum_{j>i} n\, a_{ij}\, r_{ij}^{(n-2)} [(\mathbf{r}_i - \mathbf{r}_j) \cdot \mathbf{r}_i + (\mathbf{r}_j - \mathbf{r}_i) \cdot \mathbf{r}_j] \quad \text{N1.1.1}$$

$$= -n \sum_i \sum_{j>i} a_{ij}\, r_{ij}^{(n-2)} [(\mathbf{r}_i - \mathbf{r}_j) \cdot (\mathbf{r}_i - \mathbf{r}_j)] = -n \sum_i \sum_{j>i} a_{ij}\, r_{ij}^{n-2}\, r_{ij}^{2}.$$

Thus,

$$\sum_i \mathbf{f}_i \cdot \mathbf{r}_i = -n \sum_i \sum_{j>i} a_{ij}\, r_{ij}^{n} = -n \sum_i \sum_{j>i} \Phi(r_{ij}). \quad \text{N1.1.2}$$

Since the second summation is only over $j > i$, there is no "double-counting" involved, and the double sum is just the total potential energy of the system.

1.2 As in Section 2, let us assume that the force density is derivable from a potential which is a homogeneous function of the distance between the source and field point.[5] Then, we can write the potential as

$$\Phi(\mathbf{r}) = \int_{V'} \rho(\mathbf{r}')\, (|\mathbf{r} - \mathbf{r}'|)^n\, dV' \quad n < 0 \quad \text{N1.2.1}$$

and the force density is then

$$\mathbf{f}(\mathbf{r}) = -\rho(\mathbf{r})\, \nabla_r \Phi(\mathbf{r}) = -\rho(\mathbf{r}) \int_{V'} \rho(\mathbf{r}')\, \nabla_r (|\mathbf{r} - \mathbf{r}'|)^n\, dV. \quad \text{N1.2.2}$$

while the force density at a source point due to all the field points is

$$\mathbf{f}(\mathbf{r}') = -\rho(\mathbf{r}')\, \nabla_{r'} \Phi(\mathbf{r}') = -\rho(\mathbf{r}') \int_{V} \rho(\mathbf{r}')\, \nabla_{r'} (|\mathbf{r} - \mathbf{r}'|)^n\, dV \quad \text{N1.2.3}$$

where ∇_r and $\nabla_{r'}$ denote the gradient operator evaluated at the field point $\mathbf{r}$ and the source point $\mathbf{r}'$ respectively. Since the contribution to the force density from any pair of sources and field points will lie along the line joining the two points,

$$\nabla_r (|\mathbf{r} - \mathbf{r}'|)^n = -\nabla_{r'} (|\mathbf{r} - \mathbf{r}'|)^n = n(|\mathbf{r} - \mathbf{r}'|)^{n-2} (\mathbf{r} - \mathbf{r}'). \quad \text{N1.2.4}$$

Now $\int_V \mathbf{f}(\mathbf{r}) \cdot \mathbf{r}\, dV = \int_{V'} \mathbf{f}(\mathbf{r}) \cdot \mathbf{r}'\, dV$ so multiplying N 1.2.2 by $\mathbf{r}$ and integrating over V produces the same result as multiplying N 1.2.3 by $\mathbf{r}'$ and integrating over V'. Thus, doing this and adding N 1.2.2 to N 1.2.3, we get

$$2 \int_V \mathbf{f} \cdot \mathbf{r}\, dV = -\int_V \rho(\mathbf{r}) \int_{V'} \rho(\mathbf{r}')\, \mathbf{r} \cdot \nabla_r (|\mathbf{r} - \mathbf{r}'|)^n\, dV'\, dV \quad \text{N1.2.5}$$

$$= \int_{V'} \rho(\mathbf{r}') \int_V \rho(\mathbf{r})\, \mathbf{r} \cdot \nabla_{r'} (|\mathbf{r} - \mathbf{r}'|)^n\, dV\, dV'.$$

1.3 It should be noted that the left hand side of 1.5.2 is zero if the system is periodic and the integral is taken over the period.

References

1. Kurth, Rudolf (1957), Introduction to the Mechanics of Stellar Systems, Pergamon Press, N. Y., London, Paris, p. 69.
2. Landau, L. D. & Lifshitz, E. M. (1962), The Classical Theory of Fields, Trans. M. Hamermesh, Addison-Wesley Pub. Co., Reading, Mass. USA, p. 95-97.
3. Chandrasekhar, S. (1957), An Introduction to the Study of Stellar Structure, Dover Pub., Inc., p. 49-51.
4. Goldstein, H. (1959), Classical Mechanics, Addison-Wesley Pub. Co., Reading, Mass., p. 69-71.
5. Landau, L. D. & Lifshitz, E. M. (1960), Mechanics, Trans. J. B. Sykes & J. S. Bell, Addison-Wesley Publ. Co., Reading, Mass., p. 22-23.
6. Boltzmann, L. (1811), Sitzler Akad. Wiss. Wien 63, 397, 679.
7. Maxwell, J. C. (1879), Trans. Cam. Phil. Soc. 12, 547.
8. Rosenthal, A. (1913), Ann. der Physik 42, 796.
9. Plancherel, M. (1913), Ann. der Physik 42, 1061.
10. Ogorodnikov, K. F. (1965), Dynamics of Stellar Systems, Pergamon Press, The McMillan Co., New York, p. 153.
11. Farquhar, I. E. (1964), Ergodic Theory in Statistical Mechanics, Interscience Pub., John Wiley & Sons, Ltd., London, New York, Sydney, p. 77.
12. _____. ibid., p. 3.
13. Birkhoff, G. D. (1931), Proc. of Nat'l. Acad. of Sci.: U.S. 17, p. 656.
14. Lewis, R. M. (1966), Arch. Rational Mech. Anal 5, p. 355.
15. Sinai, Ya. (1962), Vestnik Mossovskova Gosudrastvennova Universitata Series Math. 5.
16. Arnold, V. I., and Avez, A. (1968), Ergodic Problem of Classical Mechanics, W. A. Benjamin, Inc., New York, Amsterdam, p. 78.
17. Farquhar, I. E. (1964), Loc. cit. p. 23-32.

2. Contemporary Aspects of the Virial Theorem

1. The Tensor Virial Theorem

The tensor representation of the virial theorem is an attempt to restore some of the information lost in reducing the full vector equations of motion described in Section 1 to scalars. Although the germ of this idea can be found developing as early as 1903 in the work of Lord Rayleigh[1], it wasn't until the 1950's that Parker[2,3], and Chandrasekhar and Fermi[4] found the concept particularly helpful in dealing with the presence of magnetic fields. The concept was further expanded by Lebovitz[5] and in a series of papers by Chandrasekhar and Lebovitz[6,7] during the 1960's, for the investigation of the stability of various gaseous configurations. However, the most lucid derivation is probably that presented by Chandrasekhar in 1961[8] and it is a simplified version of that derivation which I shall give here.

As previously mentioned the motivation for this approach is to regain some of the information lost in forming the scalar virial theorem by keeping track of certain aspects of the system associated with its spatial symmetries. If one recalls the full-blown vector equations of motion in Chapter 1, Section 1, this amounts to keeping some of the component information of those equations, but not all. In particular, it is not surprising that since system symmetries inspire this approach that the information to be kept relates to motions along orthogonal coordinate axes.

At this point, it is worth pointing out that the derivation in Chapter 1, Section 2, essentially originates from the equations of motion of the system being considered. The derivations take the form of multiplying those equations of motions by position vectors and averaging over the spatial volume. The final step involves a further average over time. That is to say that the virial theorem results from taking spatial moments of the equations of motion and investigating their temporal behavior. (Recall that the equations of motion themselves are moments of the Boltzmann transport equation.) Since moment analysis of this type also yields some of the most fundamental conservation laws of physics (i.e., momentum, mass and energy), it is not surprising that the virial theorem should have the same power and generality of these laws. Indeed, it is rather satisfying to one who believes that "all that is good and beautiful in physics" can be obtained from the Boltzmann equation, that the virial theorem essentially arises from taking higher order moments of that equation.

With that in mind let us consider a collisionless pressure-free system analogous with that considered in Chapter 1, Section 2 and neglect viscous forces and macroscopic forces such as net rotation and magnetic fields as we shall consider them later. Under these conditions, equation 1.1.4 becomes

$$\rho \frac{\partial \mathbf{u}}{\partial t} + \rho (\mathbf{u} \cdot \nabla) \mathbf{u} = -\rho \nabla \Phi = \rho \frac{d\mathbf{u}}{dt} \qquad 2.1.1$$

which is simply the vector representation of either equations (1.1.4) or (1.2.1). In Chapter 1, Section 2, we essentially took the inner product of 1.1.1 with the position vector **r** and integrated over the volume to produce a scalar equation. Here we propose to take the outer product of equation 1.1.1 with the position vector **r** producing a tensor equation which can be regarded as a set of equations relating the various components of the resulting tensors. Cursory dimensional arguments should persuade one that this procedure should produce relationships between the various moments of inertia of the system and energy-like tensors. Thus, our starting point is

$$\int_V \rho \mathbf{r} \frac{d\mathbf{u}}{dt} dV = \int_V \rho \mathbf{r} \nabla\Phi dV . \qquad 2.1.2$$

Although some authors choose a slightly different convention, the term on the right hand side of 2.1.2 can properly be called the virial tensor.

Now, as before let the potential be

$$\Phi(\mathbf{r}) = \int_V \rho(\mathbf{r}') (|\mathbf{r} - \mathbf{r}'|)^n dV \quad n<0 \qquad 2.1.3$$

Then following exactly the same manipulation as in Chapter 1, only taking into account outer products instead of inner products with the position vectors, we get the virial tensor.[2.1]

$$\int_V \rho \mathbf{r} \nabla\Phi dV = n \{ \frac{1}{2} \int\int \rho(\mathbf{r}) \rho(\mathbf{r}')$$
$$\cdot [\mathbf{r} - \mathbf{r}'] [\mathbf{r} - \mathbf{r}'](|\mathbf{r} - \mathbf{r}'|)^{n-2} dV' dV \} \qquad 2.1.4$$

If we define

$$\mathbf{I} = \int_V (\rho \mathbf{r}\mathbf{r}) dV$$

$$\mathbf{T} = \frac{1}{2} \int_V (\rho \mathbf{u}\mathbf{u}) dV$$

$$\mathbf{U} = \frac{1}{2} \int_V \int_{V'} \rho(\mathbf{r}) \rho(\mathbf{r}) [\mathbf{r} - \mathbf{r}'] [\mathbf{r} - \mathbf{r}']$$
$$\cdot (|\mathbf{r} - \mathbf{r}'|)^{n-2} dV' dV . \qquad 2.1.5$$

equation 2.1.2 becomes

$$\frac{1}{2} \frac{d^2 \mathbf{I}}{dt^2} = 2\mathbf{T} + n\mathbf{U} \qquad 2.1.6$$

which is essentially the tensor representation of Lagrange's identity[2.2] where I is sometimes called the moment of inertia tensor, T the kinetic energy tensor and U the potential energy tensor.

By eliminating additional external forces such as magnetic fields and rotation we have lost much of the power of the tensor approach. However, some insight into this power can be seen by considering in component form one term in the expansion of the virial tensor (See note 2.1).

$$\int_V \rho \frac{d}{dt}\left(\mathbf{r}\frac{d\mathbf{r}}{dt}\right) dV = \int_V \rho \frac{d}{dt}\left(x_i \frac{dx_i}{dt}\right) dV . \qquad 2.1.7$$

Since this tensor is clearly symmetric we find, by using the same conservation of mass arguments discussed earlier, that

$$\frac{d}{dt}\int_V \rho\left(x_i \frac{dx_i}{dt} - x_j \frac{dx_i}{dt}\right) dV = 0 \qquad 2.1.8$$

which simply says the angular momentum about x_k is conserved. Thus the tensor virial theorem leads us to a fundamental conservation law which would not have been apparent from the scalar form derived earlier.

2. Higher Order Virial Equations

In the last section it became clear that both the scalar and tensor forms of the virial theorem are obtained by taking spatial moments of the equations of motion. Chandrasekhar[9] was apparently the first to note this and to inquire into the utility of taking higher moments of the equations of motion. There certainly is considerable precedent for this in mathematical physics. As already noted, moments in momentum space of the Boltzmann transport equation yield expressions for the conservation of mass, momentum and energy. Spatial moments of the transport equation of a photon gas can be used to obtain the equation of radiative transfer. Approximate solutions to the resulting equations can be found if suitable assumptions such as the existance of an equation of state are made to "close" the moment equations. Such is the origin of such diverse expressions as the Eddington approximation in radiative transfer, the diffusion approximation in radiative transfer, the diffusion approximation in gas dynamics and many others. Usually, the higher the order of the moment expressions, the less transparent the physical content. Nevertheless, in the spirit of generality, Chandrasekhar investigated the properties of the first several moment equations. In a series of papers, Chandrasekhar and Lebovitz[10,11] and later Chandrasekhar[12,13] developed these expressions as far as the fourth-order moments of the equations of motion.

Since for no moment expressions other than the first moment do any terms ever appear that can be identified with the Virial of Claussius, it is arguable as to whether they should be called virial expressions at all. However, since it is clear that this investigation was inspired by studies of the classical virial theorem, I will briefly review their development. Recall the Euler-Lagrange equation of hydrodynamic flow developed in Chapter 1, equation 1.1.4

$$\frac{\partial \mathbf{u}}{\partial t} + (\mathbf{u} \cdot \nabla)\mathbf{u} = -\nabla\Phi - \frac{1}{\rho}\nabla \cdot \mathbf{P} - \frac{1}{\rho}\int_V \mathbf{s}(\mathbf{v}-\mathbf{u})dV \qquad 2.2.1$$

Quite simply the nth order "virial equations" of Chandrasekhar are generated by taking (n-1) outer tensor products of the radius vector **r** and equation 2.2.1. The result is then integrated over all physical space. This leads to a set of tensor equations containing tensors of rank n. If we assume that particle collisions are isotropic, then the source term of 2.2.1 vanishes and $\nabla \cdot \mathbf{P} = \nabla P$. The symbolic representation for the nth order "virial equation" can then be written as:

$$\int_V \mathbf{r}^{(n-1)}\rho \frac{d\mathbf{u}}{dt} dV + \int_V \rho \mathbf{r}^{(n-1)} \nabla\Phi dV$$

$$+ \int_V \mathbf{r}^{(n-1)} \nabla P dV = 0 \qquad 2.2.2$$

Recalling our arguments in Chapter 1, section 4, about conservation of mass, it follows from equation 1.4.6 that

$$\frac{d}{dt}\left(\int_V \rho Q dV\right) = \int_V \rho \frac{dQ}{dt} dV , \qquad 2.2.3$$

where Q is any point defined property of the medium. Thus, the first order "virial equations", just correspond to the integrals of the equation of motion themselves over volume. So

$$\frac{d}{dt} \int_V \rho \mathbf{u} \, dV + \int_V \rho \nabla \phi \, dV + \int_V \nabla P \, dV = 0 \quad . \qquad 2.2.4$$

Noting that $\nabla Q = \nabla \cdot (\mathbb{1} Q/3)$, the second and third integrals are zero by the divergence theorem and 2.2.4 becomes

$$\frac{1}{2} \frac{d^2}{dt^2} \left(\int_V \rho \mathbf{r} \, dV \right) = 0 \qquad 2.2.5$$

essentially telling us that the center of mass ($\int_V \rho \mathbf{r} \, dV$) is not being accelerated. Setting n = 2 we arrive at the second order "virial equations" which are just the tensor version of Lagrange's identity which we discussed in the previous section.

In general, we can represent the nth order "virial equations" as

$$\int_V \rho \mathbf{r}^{(n-1)} \frac{d\mathbf{u}}{dt} \, dV + \int_V \rho \mathbf{r}^{(n-1)} \nabla \phi \, dV$$
$$+ \int_V \mathbf{r}^{(n-1)} \nabla P \, dV = 0 \qquad 2.2.6$$

which after use of continuity and the divergence theorem becomes

$$\frac{d}{dt} \int_V \rho \mathbf{r}^{(n-1)} \mathbf{u} \, dV - \int_V \rho \frac{d}{dt}(\mathbf{r}^{(n-1)}) \mathbf{u} \, dV$$

$$+ \int_V \rho \mathbf{r}^{(n-1)} \nabla \phi \, dV + \int_V P \mathbb{1} \cdot \nabla [\mathbf{r}^{(n-1)}] \, dV = 0 \quad . \qquad 2.2.7$$

Since the outer product in general does not commute, the integrals of the second and fourth term become strings of tensors of the form

$$\rho \frac{d}{dt}[\mathbf{r}^{(n-1)}]\mathbf{u} = \mathbf{u}[\mathbf{r}^{(n-2)}]\mathbf{u} + \mathbf{r}\,\mathbf{u}[\mathbf{r}^{(n-3)}]\mathbf{u}$$
$$+ \ldots + \mathbf{r}^{(n-2)} \mathbf{u}\,\mathbf{u} \qquad 2.2.8$$

and $\quad P\nabla(\mathbf{r}^{(n-1)}) = P[\mathbb{1}\,\mathbf{r}^{(n-2)} + \mathbf{r}\,\mathbb{1}\,\mathbf{r}^{(n-3)}$
$$+ \ldots + \mathbf{r}^{(n-2)} \mathbb{1}\,]$$

However, commutivity does not apply to the first term to yield

$$\frac{d}{dt} \int_V \rho \mathbf{r}^{(n-1)} \mathbf{u} \, dV = \frac{1}{n!} \frac{d^2}{dt^2} \int_V \rho \mathbf{r}^{(n)} dV \qquad 2.2.9$$

Thus, each term in 2.2.7 represents one or more tensors of rank n, the first of which is the second time derivative of a generalized moment of inertia tensor and the last three are all 'energy-like' tensors. From equation 2.2.8, it is clear that the second integral will generate tensors which are spatial moments of the kinetic energy distributions while the last term will produce moment tensors of the pressure distribution. The third integral is, however, the most difficult to rigorously represent. For n = 2 we know it is just the total potential energy. Chandrasekhar[8] shows how these tensors can be built up from the generalized Newtonian tensor potential or alternatively from a series of scalar potentials which obey the equations

$$\nabla^2 \psi = -4\pi G \rho$$

$$\nabla^4 \chi = -8\pi G \rho$$

$$\nabla^6 \mathcal{K} = -32\pi G \rho$$
$$\vdots$$

$$2.2.10$$

Thus, we have formulated a representation of what Chandrasekhar calls the "higher order virial equations". They are, in fact, spatial tensor moments of the equations of motion. We may expect them to be of importance in the same general way as the virial theorem itself. That is, in stationary systems the left hand side of equation 2.2.7 vanishes and the result is a system of identities between the various tensor energy moments. Keeping in mind that any continuum function can be represented in terms of a moment expansion, equations 2.2.7 must thus contain all of the information concerning the structure of the system. These equations thus represent an alternate form to the solution of the equation of motions. Like most series expansions, it is devoutly to be wished that they will converge rapidly and the "higher order" tensors can usually be neglected.

3. Special Relativity and the Virial Theorem

So far we have considered only the virial theorem that one obtains from the Newtonian equations of motion. Since there are systems such as white dwarfs, wherein the dynamic pressure balancing gravity is supplied by particles whose energies are very much larger than their rest energy, it is appropriate that we investigate the extent to which we shall have to modify the virial theotem to include the effects of special relativity.

For systems in equilibrium, the virial theorem says $2T = \Omega$. One might say that it requires a potential energy equal to 2T to confine the motions of particles having a total kinetic energy T. As particles approach the velocity of light the kinetic energy increases without bound. One may interpret this as resulting from an unbounded increase of the particle's mass. This increase will also affect the gravitational potential energy, but the effect is quadratic. Thus we might expect in a relativistic system that a potential energy less than 2T would be required to maintain equilibrium. This appears to be the conclusion arrived at by Chandrasekhar when, while investigating the internal energy of white dwarfs he concludes that as the system becomes more relativistically degenerate, T approaches Ω and this "must be the statement of the virial theorem for material particles moving with very nearly the velocity of light."[14] This is indeed the case and is the asymtotic limit represented by a photon gas or polytrope of index 3.

In order to obtain the somewhat more general result of a relativistic form of Lagrange's identity, we shall turn to the discussion of relativistic mechanics of Landau and Lifshitz[15] mentioned in Chapter 1. As most discussions in field mechanics generally start from a somewhat different prospective, let us examine the correspondence with the starting point of the equations of motion adopted in the earlier sections.

Generally most expositions of field mechanics start with the statement that

$$\Box \cdot \mathfrak{J} = 0 \qquad \qquad 2.3.1$$

where $\mathfrak{J}$ is the Maxwell stress-energy tensor.

This is equivalent to saying that there exists a volume in space-time sufficiently large so that outside that volume the stress-energy tensor is zero. This equivalence is made obvious by applying Gauss' divergence theorem so that

$$\int_R \Box \cdot \mathfrak{J} \, dR = \int_S \mathfrak{J} \cdot d\mathbf{s} = 0 \quad . \qquad 2.3.2$$

In short, 2.3.1 is a conservation law. We have already seen that the fundamental conservation laws of physics are derivable from the Boltzmann transport equation as are the equations of motion. Indeed, the operation of taking moments is quite similar in both cases. Thus, both starting points are equivalent as they have their origin in a common concept.

Although the conceptual development for this derivation is inspired by Landau and Lifshitz the subscript notation will be largely that employed by Misner, Thorne, and Wheeler[16]. Tempting as it is to use the coordinate-free geometry of these authors, the concept of taking moments at this point is most easily understood within the context of a coordinate representation so for the moment we will keep that approach.

In a Lorentz coordinate system, Landau and Lifshitz give components of the 4-velocity of a particle as

$$u_\alpha = \frac{dx_\alpha}{ds} \qquad \alpha = 0 \ldots 3 \qquad 2.3.3$$

where $\frac{ds}{dt} = c(1-v^2/c^2)^{1/2} = \gamma c$

The components of the energy-momentum tensor are

$$\mathfrak{J}_{\alpha\beta} = \rho c \, u_\alpha u_\beta \, ds/dt \qquad 2.3.4$$

and specifically $\mathfrak{J}_{oj} = i\rho c^2 u_j$

which are clearly symmetric in α and β. Since $\sum_{\alpha=0}^{3} u_\alpha^2 = -1$, the trace of 2.3.4 is just

$$\sum_{\alpha=0} \mathfrak{J}_{\alpha\alpha} = -\rho c^2 \gamma \qquad 2.3.5$$

In terms of the three-dimensional components the conservation law expressed by 2.3.1 can be written as

$$\frac{1}{ic}\frac{\partial \mathfrak{J}_{jo}}{\partial t} + \sum_{k=1}^{3} \frac{\partial \mathfrak{J}_{jk}}{\partial x_k} = c\frac{\partial(\rho u_j)}{\partial t} + \sum_{k=1}^{3} \frac{\partial \mathfrak{J}_{jk}}{\partial x_k} = 0 \qquad 2.3.6$$

Substituting for the components of $\mathfrak{J}$ and repeating the algebra of earlier derivations we have[2,3]

$$\sum_{j=1} \frac{d}{dt} \int_V \frac{\rho}{2} \frac{d}{dt}(x_j x_j/\gamma) \, dV = \Omega + T + \int_V \gamma\tau \, dV . \qquad 2.3.7$$

Again, using the conservation of mass arguments in Chapter 1, this becomes

$$\sum_{j=1} \frac{1}{2} \frac{d^2}{dt^2} \int_V (\rho x_j x_j/\gamma) \, dV = \Omega + T + \int_V \gamma\tau \, dV. \qquad 2.3.8$$

where, if we define the volume integral on the left to be the relativistic moment of inertia, we can write

$$\frac{1}{2} \frac{d^2}{dt^2}\left(I_r\right) = \Omega + T + \int_V \gamma\tau \, dV \qquad 2.3.9$$

In the low velocity limit $\gamma \to 1$ so that $I_r \to I$ and we recover the ordinary Lagrange's identity. In the relativistic limit as $\gamma \to 0$ we recover for stable systems the Chandrasekhar result that $T + \Omega = 0$ (i.e., the total energy $E = 0$). Thus, it is fair to say that equation (2.3.9) is an expression of the Lagrange's identity including the effects of special relativity.

It is worth noting by analogy with Section 1 that the tensor relativistic theorem can be derived by taking the outer or 'tensor' product of the space-like position vector with equation 2.3.6 and integrating over the volume. Following the same steps that lead to equation N.2.3.3, we get

$$c \int_V x_\ell \frac{\partial(\rho u_j)}{\partial t} \, dV - \int_V \mathfrak{J}_{\ell j} \, dV = 0 . \qquad 2.3.10$$

Since $\mathfrak{J}_{\ell j}$ is symmetric, we can add 2.3.10 to its counterpart with the indices interchanged, and get

$$c \int_V [x_\ell \frac{\partial(\rho u_j)}{\partial t} + x_j \frac{\partial(\rho u_\ell)}{\partial t}] \, dV - 2 \int_V \mathfrak{J}_{\ell j} \, dV = 0 \qquad 2.3.11$$

or finally

$$\frac{1}{2} \frac{d^2}{dt^2} \int_V [\rho(x_j x_\ell)/\gamma] \, dV = \int_V \mathfrak{J}_{\ell j} \, dV . \qquad 2.3.12$$

The integral on the left can be viewed on the relativistic moment of inertia tensor while the right hand side is just the volume integral of the components of the energy momentum tensor. Following the prescription used to generate (2.3.11), but subtracting the transpose of 2.3.11 from itself, yields

$$\int_V [x_\ell \frac{\partial(\rho u_j)}{\partial t} - x_j \frac{\partial(\rho u_\ell)}{\partial t}] \, dV = 0, \qquad 2.3.13$$

which becomes by application of Liebnitz's law

$$\frac{d}{dt}\left[\int_V \frac{\rho}{\gamma} \{ x_\ell \frac{dx_j}{dt} - x_j \frac{dx_\ell}{dt} \} \, dV\right] = 0 \qquad 2.3.14$$

and is the relativistic form of the expression for the conservation of angular momentum obtained in Section 1, equation 2.1.8.

4. General Relativity and the Virial Theorem

The development of quantum theory and the formulation of the general theory of relativity probably represent the two most significant advances in physical science in the first half of the twentieth century. In light of the general nature and wide applicability of the virial theorem it is surprising that little attempt was made during that time to formulate it within the context of general relativity. Perhaps this was a result of the lack of physical phenomena requiring general relativity for their description or possibly the direction of mathematical development undertaken for theory itself. For the last twenty years, there has been a concerted effort on the part of relativists to seek coordinate-free descriptions of general relativity in order to emphasize the connection between the fundamental geometrical properties of space and the description of associated physical phenomena. Although this has undoubtedly been profitable for the development of general relativity, it has drawn attention away from that technique in theoretical physics known as 'moment analysis'. This technique produces results which are in principle coordinate independent but usually utilize some specific coordinate frames for the purpose of calculation.

Another point of difficulty consists of the nature of the theory itself. General relativity, like so many successful theories, is a field theory and is thus concerned with functions defined at a point. Virtually every version of the virial theorem emphasizes its global nature.* That is, some sort of symmetrical volume is integrated or summed over to produce the appropriate physical parameters. This difference becomes a serious problem when one attempts to assign a physical operational interpretation to the quantities represented by the spatial integrals. The problem of operational definition of macroscopic properties in general relativity has plagued the theory since its formulation. Although continuous progress is being made, there does not exist any completely general formulation of the virial theorem within the framework of general relativity at this time. This certainly is not to say that such a formulation cannot be made. Indeed, what we have seen so far should convince even the greatest skeptic that such a formulation does exist since its origin is basically that of a conservation law. Even the general theory of relativity recognizes conservation laws although their form is often altered.

Let us take a closer look at the origin of some of these problems. This can be done by taking into account in a self-consistent manner in the

---

* It is worth noting that in order to 'test' any field theory against observation, it is necessary to compare integrals of the field quantities with the observed quantities. Even something as elementary as density is always "observed" by comparing some mass to some volume. It is impossible in principle to measure anything at a point. This obvious statement causes no trouble as long as we are dealing with concepts well within the range of our experience where we can expect our intuition to behave properly. However, beyond this comfortable realm, we are liable to attribute physical significance and testability to quantities which are in principle untestable. The result is to restrict the range of possibility for a theory unnecessarily.

Einstein field equations all terms of order $1/c^2$. This is the approach adopted by Einstein, Infield, and Hoffman[19] in their approach to the relativistic n-body problem and successfully applied by Chandrasekhar[20] to hydrodynamics. Although more efficient approximation techniques exist for the calculation of higher order relativistic phenomena such as gravitational radiation, this time honored approach is adequate for calculating the first order (i.e. c^2) terms commonly known as the post-Newtonian terms.**

During the first half of 1960, in studying the hydrodynamics of various fluid bodies, Chandrasekhar developed the virial theorem to an extremely sophisticated level. The most comprehensive recognition of this work can be found in his excellent book on the subject[9]. One of the highlights not dealt with in the book are his efforts to include the first-order effects of general relativity. In an impressive and lengthy paper during 1965, Chandrasekhar developed the post-Newtonian equation of hydrodynamics including a formulation of the virial theorem[20]. It is largely this effort which we shall summarize here.

One of the fundamental difficulties with the general theory of relativity is that it is a non-linear theory. The physical properties of matter are represented by the geometry of space and in turn determines the geometry of space. It is this non-linearity that causes so much difficulty with approximation theory and with which the Einstein, Infield, Hoffman theory (EIH) deals directly. The basic approach assumes that the metric tensor can be written as being perturbed slightly from the flat-space or Euclidean metric so that

$$g_{\alpha\beta} = g^{(o)}_{\alpha\beta} + h_{\alpha\beta} \qquad 2.4.1$$

** For a beautifully concise and complete summary of the post-Newtonian approximation, see Misner, Thorne, and Wheeler, <u>Gravitation</u> (1973), W. H. Freeman & Co., San Francisco, Chapter 39.

where the $\mathfrak{h}_{\alpha\beta}$ are small terms of the order of $1/c^2$ or smaller. This enables one to determine the elements of the energy-momentum tensor up to terms of the order of $1/c^2$ from its definition

$$\mathfrak{J}_{\alpha\beta} = (\varepsilon + P) u_\alpha u_\beta - P \mathfrak{g}_{\alpha\beta} \qquad 2.4.2$$

The Einstein field equations can be written in terms of the Ricci tensor and the energy-momentum tensor as

$$\mathfrak{R}_{\alpha\beta} = - \frac{8\pi G}{c^4} (\mathfrak{J}_{\alpha\beta} - \frac{1}{2} J \mathfrak{g}_{\alpha\beta}) \qquad 2.4.3$$

where J is just the trace of $\mathfrak{J}_{\alpha\beta}$. The Ricci tensor is essentially a geometric tensor and contains information relating to the metric alone. The EIH approximation provides a prescription for solving the field equations in various powers of $1/c^2$ given the information concerning $\mathfrak{J}_{\alpha\beta}$ and $\mathfrak{g}_{\alpha\beta}$. In general the procedure determines the metric coefficients to one higher order than was originally specified. This procedure can be repeated but there remain some unsolved problems as to convergence of the scheme in general. At any point one may use the perturbed metric and the prescription for obtaining the equations of motion to generate a set of perturbed equations of motion.

The relativistic prescription that free particles follow geodesic paths is logically equivalent to stating the four-space divergence of the energy-momentum tensor is zero. That is

$$\Box \cdot \mathfrak{J}_{\alpha\beta} = 0 . \qquad 2.4.4$$

Indeed it is this condition that in the flat-space metric yields of Euler-Lagrange's equations of hydrodynamic flow. It was this condition that we needed in Section 3 to obtain a form of the virial theorem appropriate for special relativity. Unfortunately the process of taking the divergence loses one order of approximation and thus it is not possible to go directly from the perturbed metric to the equations of motion and maintain the same level of accuracy. One must first pass through the field equations and the EIH approximation scheme.

In order to follow this prescription one must first start with an approximation to the metric $\mathfrak{g}_{\alpha\beta}$. Here it is traditional to invoke the principle of equivalence that requires that[21]

$$h_{\alpha\beta} = -\frac{2\Psi}{c^2}\delta_{\alpha\beta} + O(< 1/c^3) \qquad 2.4.5$$

With this as a starting point one may proceed with the approximation scheme and obtain the equations of motion. Like many approximation schemes the mathematical manipulations are formidable and physical insight easily lost. However, progress is rapid in a field such as this and what was an original research effort by Chandrasekhar in 1965 becomes a 'homework' problem for Misner, Thorne and Wheeler in 1972 (M.T.W. exercise 39.13.) Although it is contrary to the spirit of this book to quote results without derivation, I find in this case I must. We have laid neither a sufficient mathematical framework nor developed the general theory of relativity sufficiently to present the derivation in detail without consuming excessive space. Instead, let us consider the types of effects we might expect general relativity to introduce and see if these can be identified in the resulting equations of motion.

Firstly, energy is matter and therefore its motion must be followed in the equations of motion as well as that of matter. This is really a consequence of special relativity but insofar as this 'added' mass affects the metric, we should find its effects present. The distortion of space also changes or at least complicates what is meant by a volume and thus it is useful to define a density ρ^* which obeys a continuity equation

$$\frac{\partial \rho^*}{\partial t} + \nabla \cdot (\rho^* \mathbf{u}) = 0 \qquad 2.4.6$$

where
$$\rho^* = \rho_o[1 + \frac{1}{c^2}(u^2 + 3\Psi)]$$

For purposes of simplification, Chandrasekhar finds it convenient to define a slightly altered form of the density which explictly contains the internal energy of the gas and has a slightly altered continuity equation

$$\frac{\partial \sigma}{\partial t} + \nabla \cdot (\sigma \mathbf{u} + \frac{1}{c^2}(\rho_o \frac{\partial \Psi}{\partial t} - \frac{\partial \rho_o}{\partial t})) = 0 \qquad 2.4.7$$

where
$$\sigma = \rho_o[1 + \frac{1}{c^2}(u^2 + 2\Psi + \Pi + p/\rho_o)]$$

Here Π/c^2 is the internal energy of the gas and p is the local pressure. It is worth noting that ρ_o is the density one would find in the absence of general relativity but where relativistic effects are important it is essentially a non-observable quantity since one could not devise a test for measuring it.

The general theory of relativity is a non-linear theory and thus we should expect terms to appear which reflect this non-linearity. They will be of different types. Firstly one should expect effects of the Newtonian potential Ψ affecting the metric directly which in turn modifies Ψ. These terms are indeed present but Misner, Thorne and Wheeler show that they can be represented by direct integrals over the mass distribution[22]. Secondly, since the matter and the metric are inexorably tied together, motion of matter will 'drag' the metric which will introduce velocity dependent terms in the 'potentials' used to represent those terms.

Both these effects can indeed be represented by 'potentials' but not just the Newtonian potential. Thus, various authors introduce various kinds of potentials to account for these non-linear terms. With this in mind the equations of motion as derived by Chandrasekhar become[23]

$$\frac{d}{dt}(\sigma \mathbf{u}) - \rho_o \nabla \Psi + \nabla[(1+2\Psi/c^2)P]$$
$$+ \frac{\rho_o}{c^2}\left\{\frac{d}{dt}[4\mathbf{u}\Psi - \frac{7}{2}\mathbf{Y} - \frac{1}{2}\nabla(|\mathbf{Y}|)]\right.$$
$$\left. + 4\mathbf{u}\cdot(\nabla\mathbf{Y}) - \frac{1}{2}(\mathbf{u}\cdot\nabla)[\mathbf{Y} - (|\mathbf{Y}|)]\right\}$$
$$+ \frac{1}{2\pi G c^2}[\nabla^2\Phi\nabla\Psi + \nabla^2\Psi\nabla\Phi] = 0 \quad .$$

2.4.8

Here, the various potentials which we have introduced can be defined by the fact that they satisfy a Poisson's equation of the form

$$\nabla^2 \Psi = -4\pi G \rho_o$$

$$\nabla^2 \Phi = -4\pi G \rho_o \phi$$

$$\nabla^2 \mathbf{Y} = -4\pi G \rho_o \mathbf{u}$$

2.4.9

where $\phi = u^2 + \Psi + (1/2)\Pi + (3/2)(P/\rho_o)$.

Thus **Y** is a vector potential whose source is the same as that of the Newtonian potential weighted by the local velocity field. Similarly, Φ is a scalar potential whose source is again that of the Newtonian potential but this time weighted by a function ϕ related to the total internal energy field.

Expansive as the equations of motion are we may still derive some comprehension for the meaning of the various terms in 2.4.8. The first two terms are basically Newtonian, indeed neglecting the contribution to the mass from energy $\sigma = \rho_o$, and they are identical to the first term of the Newtonian-Euler-Lagrange equations of hydrodynamic flow. The first part of the third term is just the pressure gradient and thus also to be expected on Newtonian grounds alone. The remaining contribution to the pressure gradient results from the space curvature introduced by the presence of the matter and is perhaps the most likely relativistic correction to be expected. The remaining tensors are the non-linear interaction terms alluded to earlier. The lengthy expression in braces contain the effects of the 'dragging' of the metric by the matter and the induced velocity dependent terms. The last term represents the direct effect of the matter-energy potentials on the metrics and this effect in turn propagated the potentials themselves.

Having obtained the equations of motion for the system, the procedure for obtaining the general relativistic form of Lagranges' identity is the same as we have used repeatedly in earlier sections. For simplicity, we shall compute the scalar version of Lagranges' identity by taking the inner product of equation 2.4.8 with the position vector **r**. We should expect this procedure to yield terms similar to the classical derivation but with differences introduced by differences between ρ_o, ρ^*, and σ. In addition we shall take our volumes large enough so that volume integrals of divergence vanish. In this regard it is worth noting that if volume contains the entire

43

system, then by the divergence theorem

$$\int_V \nabla A \, dV = \frac{1}{3} \int_V \nabla \cdot (\mathbb{I} A) \, dV = 0 \quad . \qquad 2.4.10$$

Thus, by integrating the equations of motion over the volume yields

$$\frac{d}{dt} \int_V \{ \sigma \mathbf{u} + \frac{\rho_o}{c^2} [4\mathbf{u}\Psi - \frac{7}{2}\mathbf{Y} - \nabla(|\mathbf{Y}|)] \} \, dV = 0$$

$$= \frac{d}{dt} \int_V \mathbf{K} \, dV \quad . \qquad 2.4.11$$

after noting that remaining integrals in the braces { } of 2.4.8 can be integrated by parts to zero. Equation 2.4.11 is a statement of conservation of linear momentum. This is a useful result for simplifying 2.4.8.

Now we are prepared to write down Lagrange's identity by letting the integral of 2.4.11 be the local linear momentum density $\mathbf{K}$ and taking the scalar product of $\mathbf{r}$ with 2.4.8.

$$\mathbf{r} \cdot \frac{d\mathbf{K}}{dt} - \rho_o \mathbf{r} \cdot \nabla [\Psi + P(1+2\Psi/c^2)]$$
$$+ \frac{4\rho}{c^2} \mathbf{r} \cdot [\mathbf{u} \cdot (\nabla \mathbf{Y})] - \frac{1}{2} \mathbf{r} \cdot [\mathbf{u} \cdot \nabla][\mathbf{Y} - \nabla |\mathbf{Y}|]$$
$$- \frac{2\rho}{c^2} (\phi \mathbf{r} \cdot \nabla \Psi + \mathbf{r} \cdot \nabla \Phi) = 0$$
$$\qquad 2.4.12$$

After multiple integration by parts and liberal use of the divergence theorem, this becomes

$$\frac{d}{dt} \int_V (\mathbf{r} \cdot \mathbf{K}) dV = 2T + \Omega + 3 \int_V (1+2\Psi/c^2) P dV \qquad 2.4.13$$
$$+ \frac{1}{c^2} [4W + <\Phi> - \frac{7}{4} \mathcal{Y} - \frac{1}{4} \mathcal{Z}] \quad ,$$

where
$$T = \frac{1}{2} \int_V \sigma u^2 \, dV$$

$$\Omega = -\frac{1}{2} \int_V \rho_o \Psi \, dV$$

$$W = \int_V \rho_o u^2 \Psi \, dV$$

$$<\Phi> = \int_V \rho_o \phi \Psi \, dV$$

and
$$\mathcal{Y} = \int_V \rho_o \mathbf{u} \cdot \mathbf{Y} \, dV \qquad 2.4.14$$

$$\mathcal{Z} = \int_V \int_{V'} \{ \rho_o \rho'_o [\mathbf{u} \cdot (\mathbf{r}-\mathbf{r}')][\mathbf{u}' \cdot (\mathbf{r}-\mathbf{r}')]/|\mathbf{r}-\mathbf{r}'|^3 \} \, dV' dV \, .$$

This can be made somewhat more familiar if we re-write the left hand side of 2.4.13 so that

$$\frac{1}{2} \frac{d^2}{dt^2} \left(\int_V \sigma r^2 dV \right) + \frac{1}{c^2} \frac{d}{dt} \int_V \rho_o \left[4\Psi (\mathbf{r} \cdot \mathbf{u}) \right.$$

$$\left. - \frac{7}{2} \mathbf{r} \cdot \mathbf{Y} - \mathbf{r} \cdot \nabla (|\mathbf{Y}|) \right] dV + 2T + \Omega$$

$$+ 3 \int_V (1+2\Psi/c^2) P dV$$

$$+ \frac{1}{c^2} [4W + <\Phi> - \frac{7}{4} \mathcal{Y} - \frac{1}{4} \mathcal{Z}] \qquad 2.4.15$$

The first term on the left hand side of 2.4.15 is just $\frac{1}{2} d^2 I/dt^2$ in the Newtonian limit. The remaining terms arise from the correction to the metric resulting from the potential and the 'dragging' of the metric due to internal motion. The first two terms on the right are just what one would expect in the Newtonian limit while the next term can be related to the total internal energy. This term contains a relativistic correction resulting directly from the change in metric due to the presence of matter. The remaining terms are all energy like and the first two (W and $<\Phi>$) represent relativistic corrections arising from the change in the potential caused by the metric modification by the potential itself. The last two are metric dragging terms.

We have gone to some length to show the problems injected into the virial theorem by the non-linear aspects of general relativity. Writing Lagrange's identity as in equation 2.4.15 emphasizes the origin of the various terms - whether they be Newtonian or Relativistic. Although terms to this order should be sufficient to describe most phenomena in stellar astrophysics, we can ask if higher order terms or other metric theories of gravity provide any significant corrections to the virial theorem. The Einstein, Infield Hoffman approximation has been iterated up to 2 1/2 times [24,25] in a true tour-de-force by Chandrasekhar and co-workers looking for additional effects. At the 2 1/2 level of approximation, radiation reaction terms appear which could be significant for non-spherical collapsed objects which exist over long periods of time. Using a parameterized version of the post-Newtonian approximation[26] Ni has developed a set of hydrodynamic equations which must hold in nearly all metric theories of gravity which depend on the values (near unity) of a set of dimensionless parameters. This latter effort is useful for relating various terms in the equations to the fundamental assumptions made by different theories. Perhaps the most obvious lesson to be learned from the EIH approach to this problem is that continued application of the theory is not the way to approach the general results. However, of some consequence is the result that conservation laws for energy, momentum, and angular momentum exist and are subject to an operational interpretation at all levels of approximation. Thus it seems reasonable to conjecture that these laws as well as the virial theorem remained well posed in the general theory.

5. Complications – Magnetic Fields, Internal Energy, and Rotation

The full power and utility of the virial theorem does not really become apparent until one realizes that we need not have been particularly specific about the exact nature of the potential and kinetic energies that appear in the earlier derivation. Thus the presence of complicating forces can be included insofar as they are derivable from a potential. Similarly as long as the total kinetic energy can be expressed in terms of energies arising from macroscopic motions and internal thermal motions, it will be no trouble to express the virial theorem in terms of these more familiar parameters of the system. One may proceed in just this manner or return to the original equations of motion for the system. We shall discuss both approaches.

In Chapter 1, we derived the Euler-Lagrange equation of hydrodynamic flow. These equations of motion are completely general and are adequate to describe the effects of rotation and magnetic-fields if some care is taken with the coordinate frame and the pressure tensor $\mathcal{P}$. With this in mind, we may rewrite equation 1.1.4, noting that the left hand side is the total time derivative and that the pressure tensor can be explicitly split to include the presence of large scale electromagnetic fields. Thus

$$\frac{d\mathbf{u}}{dt} = -\nabla\Phi - \frac{1}{\rho}[\nabla \cdot (\mathcal{P}_g + \mathfrak{J})] - \frac{1}{\rho}\int \mathbf{s}(\mathbf{v}-\mathbf{u})dv \quad 2.5.1$$

where the tensor $\mathcal{P}_g$ refers to the gas pressure alone and the tensor $\mathfrak{J}$ represents the Maxwell stress tensor for the electromagnetic field which has components[27]

$$\mathfrak{J}_{ij} = E_i D_j - \frac{1}{2}\delta_{ij}\sum_k E_k D_k + H_i B_j - \frac{1}{2}\delta_{ij}\sum_k H_k B_k$$
$$2.5.2$$

or in dyadic notation

$$\mathfrak{J} = \mathbf{ED} - \frac{1}{2}\mathbf{1}[\mathbf{E}\cdot\mathbf{D}] + \mathbf{HB} - \frac{1}{2}\mathbf{1}[\mathbf{H}\cdot\mathbf{B}] \quad 2.5.3$$

For almost all cases in astrophysics, it is appropriate to ignore electrostriction and magnetostriction effects which complicate the relationships between **E** and **D** and **B** and **H**.

In the absence of these body forces, the divergence of 2.5.3 yields [2.4]

$$\nabla \cdot \mathbf{J} = \frac{1}{4\pi} \{ \mathbf{D} \rho_e - \mathbf{D} \times (\nabla \times \mathbf{D}) + c^2 [\mathbf{H} \times (\nabla \times \mathbf{H})] \} \quad 2.5.4$$

which is just the Lorentz force on the medium.

It is useful when considering a configuration in uniform rotation to transform the problem into a co-rotating coordinate frame. This enables one to see the effect of macroscopic mass motion explicitly in the formalism and thus assess its interaction with other large scale properties of the system. In addition such systematic motion is represented by the stream velocity **u** in the collision term of the Boltzmann equation making its meaning clearer. Therefore, just as we have separated the effects of the electromagnetic field from the pressure tensor, let us explicitly represent the effects of rotation.

In transforming the inertial coordinate frame to a non-inertial rotating frame attached to the system, we must allow for temporal changes in any vector seen in one frame but not the other. Goldstein[28] gives a particularly lucid account of how this is to be accomplished by use of the operator

$$\left(\frac{d}{dt}\right)_{\text{inertial}} = \left(\frac{d}{dt}\right)_{\text{non-inertial}} + \boldsymbol{w} \times \quad 2.5.5$$

where $\boldsymbol{w}$ is the angular velocity appropriate for the point-function upon which the operator acts. If we let

$$\mathbf{u} = \mathbf{w} + (\boldsymbol{w} \times \mathbf{r}) \quad 2.5.6$$

then the equations of motion for uniform rotation (i.e., $\boldsymbol{w}$ = constant) become

$$\frac{d\mathbf{w}}{dt} + 2(\boldsymbol{w} \times \mathbf{w}) + \boldsymbol{w} \times (\boldsymbol{w} \times \mathbf{r}) =$$

$$- \nabla \Phi - \frac{1}{\rho} [\nabla \cdot (\mathbf{P} + \mathbf{J})] - \frac{1}{\rho} \int \mathbf{s}(\boldsymbol{w}) \, d\mathbf{v} \quad . \quad 2.5.7$$

It can be shown that for uniform rotation the term [2.5]

$$\boldsymbol{w} \times (\boldsymbol{w} \times \mathbf{r}) = \frac{1}{2} \nabla [(\boldsymbol{w} \times \mathbf{r}) \cdot (\boldsymbol{w} \times \mathbf{r})] \, , \quad 2.5.8$$

so that it may be combined with the gravitational potential and $\frac{1}{2}(\boldsymbol{w} \times \boldsymbol{r})^2$ may be considered a 'rotational potential'. The term $2(\boldsymbol{w} \times \boldsymbol{w})$ is known as the "coriolis force". Since both the pressure tensor and Maxwell tensor are normally "fixed" to the body, we should expect their formulation in the rotating frame to be simpler. It is difficult to proceed much further with the equations of motion without making some simplifying assumptions. The most helpful and also reasonable of these is to assume that all collisions processes are isotropic in space. This has two results. Firstly, the collision term on the right hand side of the equation of motion averages to zero when integrated over all velocity space. Secondly, the gas pressure tensor $\boldsymbol{P}g$ becomes diagonal with all elements equal and can thus $\nabla \cdot \boldsymbol{P}_g = P$ where P is the scalar gas pressure. This effectively guarantees the existence of a scalar equation of state which will be useful later when relating P to the internal energy. Since almost all astrophysical situations relate to plasmas, we may consider the configurations to have a very large conductivity and can therefore neglect the contribution to the Maxwell Field tensor of electric fields. Using these assumptions and equations 2.5.4 and 2.5.8 the equation of motion becomes

$$\frac{d\boldsymbol{w}}{dt} + 2(\boldsymbol{w} \times \boldsymbol{w}) = -\nabla\left[\Phi + (\boldsymbol{w} \times \boldsymbol{r})^2\right] + \nabla P/\rho$$

$$-\frac{1}{4\pi\rho}\left[\boldsymbol{H} \times (\nabla \times \boldsymbol{H})\right]. \qquad 2.5.9$$

As we have done before we shall multiply the equations of motion by $\rho \boldsymbol{r}$, and integrate over the volume and generate a general tensor expression for Lagrange's Identity including rotation and magnetic fields.[2.6] Thus,

$$\frac{1}{2}\frac{d^2}{dt^2}(\boldsymbol{I}) + 2\boldsymbol{w} \cdot \boldsymbol{L} = 2\boldsymbol{\mathcal{T}} + n\boldsymbol{\mathcal{H}} + \boldsymbol{w} \cdot (\boldsymbol{I}\boldsymbol{w}) - \omega^2\boldsymbol{I}$$

$$-\int_V \boldsymbol{r} \nabla P dV - \frac{1}{4\pi}\int_V \boldsymbol{r}[\boldsymbol{H} \times (\nabla \times \boldsymbol{H})]dV. \qquad 2.5.10$$

where $\boldsymbol{I}$, $\boldsymbol{\mathcal{T}}$, and $\boldsymbol{\mathcal{H}}$ are moment of inertia, kinetic and potential energy tensors respectively. $\boldsymbol{L}$ is an angular momentum like tensor (see note 2.6).

In order to simplify the last two terms it is worth noting that they are derived from the divergence of two tensors. So, we may use the chain rule followed by the divergence theorem to simplify these terms. Thus

$$\int_V \mathbf{r} \nabla \cdot (\mathbf{P} + \mathbf{J})dV = \int_V \nabla \cdot [\mathbf{r}(\mathbf{P} + \mathbf{J})dV$$
$$- \int_V (\nabla \mathbf{r}) \cdot [\mathbf{P} + \mathbf{J}]dV \, . \qquad 2.5.11$$

The divergence theorem guarantees that the first integral can be written as a surface integral and if the volume is taken to be large enough can always be made to be zero. However, since magnetic fields always extend beyond the surface of what one normally considers the surface of the configuration, we will keep these terms for the moment. Thus

$$\int_V \nabla \cdot [\mathbf{r}(\mathbf{P} + \mathbf{J})] \, dV = \int_S [\mathbf{r}(\mathbf{P} + \mathbf{J})] \cdot d\mathbf{s} = \mathbf{S} \qquad 2.5.12$$

or, in terms of the components, the surface terms become

$$\mathbf{S}_{ij} = \frac{1}{4\pi} \int_S x_j (H_i \Sigma \, H_k dS_k) - \frac{1}{8\pi} \int_S x_j H^2 dS_i$$
$$- \int_S P_o x_j dS_i \, .$$

Keeping in mind that $\nabla \mathbf{r} = \mathbf{1}$ (i.e., a second rank tensor with components δ_{ij}), the second integral in 2.5.11 becomes

$$\int_V \mathbf{1} \cdot (\mathbf{P} + \mathbf{J})dV = \int_V \mathbf{1}(P + H^2/8\pi)dV + \frac{1}{4\pi} \int_V (\mathbf{H}\mathbf{H})dV \, . \qquad 2.5.13$$

Defining

$$\mathbf{\mathfrak{M}} = \frac{1}{8\pi} \int_V (\mathbf{H}\mathbf{H})dV \qquad 2.5.14$$

we arrive at the final tensor form for Lagrange's Identity, including rotation and magnetic fields.

$$\frac{1}{2}\frac{d^2 I}{dt^2} + 2\boldsymbol{w}\cdot \mathcal{L} = 2[\mathcal{T}-\mathcal{M}] + n\mathcal{H} + \boldsymbol{w}\cdot(\mathbf{I}\,\boldsymbol{w}) - \omega^2 I$$

$$+ 1\left(\int_V (P + H^2/8\pi)\,dV\right) + \mathcal{G}$$

2.5.15

where $\mathcal{L}_{kij} = \int_V \rho\,(r_k\, r_i\, \widetilde{\omega}_j - \widetilde{\omega}_k\, r_i\, r_j)\,dV$

We have suffered through the tensor derivation in order to show the complete generality of this formalism. The tensor component equations are essential for investigating non-radial oscillations and other such phenomena which cannot be represented by a simple scalar approach. However, it is easier to appreciate the physical significance of this approach by looking at the scalar counterpart of 2.5.15. In Section 1, we pointed out that the scalar form is derived by taking "inner" products of the position vector with the equation of motion while the tensor virial theorem involves "outer" or tensor products. One may either repeat the derivation of 2.5.15 taking "inner" products or 'contract' the component form of 2.5.15 over indices i and j.

The contraction of the tensors $\mathbf{I}$, $\mathcal{T}$, $\mathcal{M}$, $\mathcal{H}$ and $\mathbf{I}$ yield the moment of inertia about the origin of the coordinate system I, the kinetic energy due to internal motion T, the total magnetic energy $\mathcal{M}$, the total potential energy Φ and $\mathcal{G}$ respectively. Some care must be taken in contracting the tensors $\mathcal{L}$ and $\mathbf{I}\boldsymbol{w}$. From the definition of $\mathcal{L}$ in equation N2.6.3, it is clear that the contracted form of that expression can be written as

$$2\int_V \rho\,\mathbf{r}\cdot(\boldsymbol{w}\times\mathbf{w})\,dV = -2\,\boldsymbol{w}\cdot\int_V \rho(\mathbf{r}\times\mathbf{w})\,dV$$

$$= -2\,\boldsymbol{w}\cdot\int_V \boldsymbol{l}\,dV \qquad 2.5.16$$

where $\boldsymbol{l}$ is the net volume angular momentum on the material due to coriolis forces. We can again choose our rotating frame so that $\int_V \boldsymbol{l}\,dV$ is zero and this term must vanish from the contracted equation. The simplest method for deriving the value of the contracted form of $\boldsymbol{w}\cdot(\mathbf{I}\boldsymbol{w} - \boldsymbol{w}\mathbf{I})$ in equation 2.5.15 is again to examine the contracted form of the term giving rise to it. Since

51

$\mathbf{w} = \boldsymbol{w} \times \mathbf{r}$, we can expand the left hand side of 2.5.16 by means of identities relating to the vector triple product (see note 2.6, equation N2.6.4), and obtain

$$\int_V \rho \, \mathbf{r} \cdot [\boldsymbol{w} \times (\boldsymbol{w} \times \mathbf{r})] \, dv = \int_V \rho \, \mathbf{r} \cdot [\boldsymbol{w} \times \mathbf{v}] \, dv$$

$$= \int_V \rho [\mathbf{r} \times \boldsymbol{w}] \cdot \mathbf{v} \, dV = \int_V \rho v^2 \, dV = 2\,\mathcal{R} \qquad 2.5.17$$

where $\mathcal{R}$ is just the total energy due to rotation. Substitution of the contraction of these tensors into equation 2.5.15 yields a much simpler result

$$\frac{1}{2} \frac{d^2 I}{dt^2} = 2(T + \mathcal{R}) + \mathcal{M} + n\mathcal{U} + 3\int_V P \, dV + \sum_i \mathcal{S}_{ii} \qquad 2.5.18$$

The contraction of the surface terms yield the scalar integrals

$$\sum_i \mathcal{S}_{ii} = \frac{1}{4\pi} \int_S (\mathbf{r}_o \cdot \mathbf{H}_o)(\mathbf{H}_o \cdot d\mathbf{s})$$

$$- \int_S (P_o + H_o^2/8\pi) \, \mathbf{r}_o \cdot d\mathbf{s} \,. \qquad 2.5.19$$

Here, the subscript "o" indicates the value of the variables on the surface. In practice these integrals are usually small compared with the magnitude of the volume integrals found in the remainder of the expression.

So far, we have said little about the contribution of the volume integral of the pressure. Since even in the tensor representation this term appears as a scalar, there was no loss of generality in deferring the evaluation of the integral until now. Simple dimensional analysis will lead one to the result that the pressure integral is an 'energy-like' integral. From thermodynamic consideration, we can write the internal "heat energy" $\mathcal{U}$ as

$$\mathcal{U} = \int_V \rho c_v \mathcal{T} \, dV \qquad 2.5.20$$

where c_v is the specific heat of constant volume and is the temperature. We also know that the kinetic energy associated with the material is

$$\mathcal{J} = \frac{3}{2} Nk\mathcal{T} = \frac{3}{2} \rho(c_p - c_v)\mathcal{T} - \frac{3}{2} P \,. \qquad 2.5.21$$

Combining these two equations we get the total internal "heat energy" as

$$\mathcal{U} = \frac{2}{3} \int_V \frac{3 dV}{(c_p/c_v - 1)} = \int_V \frac{\rho\, dV}{[(c_p/c_v) - 1]} \; . \qquad 2.5.22$$

It is traditional to let $\gamma = c_p/c_v$ and if we let this be constant throughout the volume and neglect the surface term in equation 2.5.18, then we can write the scalar form of Lagrange's identity as

$$\frac{1}{2} \frac{d^2 I}{dt^2} = 2(T + \mathcal{R}) + \mathcal{M} - \Omega + 3(\gamma - 1)\mathcal{U} \; . \qquad 2.5.23$$

At the beginning of this discussion, I said that one could either derive Lagrange's identity from the equations of motion or from careful consideration of meanings of potential and kinetic energy. For example, the first and last terms on the right hand side of equation 2.5.23 are just the contribution to the total kinetic energy of the system from macroscopic motions, rotation and thermal motion respectively. The remaining terms are just a specification of the nature of the total potential energy. Thus, by realizing that $3(\gamma-1)\mathcal{U}$ is just twice the kinetic energy due to thermal motion we could have written 2.5.23 down immediately. However, it is unlikely that this could have been done for 2.5.15. Since the variational form of this equation will be useful in the next chapter, our efforts have not been wasted.

Before leaving this section on complicating phenomena, there is one last aspect to be investigated. In Chapter 1, Section 3, we noted that the inclusion of velocity dependent forces such as frictional force do not alter the results of the virial theorem as they can be averaged to zero given sufficient time. This result was apparently first noted by E. H. Milne in 1925.[29] Although no modification is made to the virial theorem some effects can be seen in Lagrange's identity and so, let us take a moment to recapitulate these arguments. If we have a volume force which can be derived from a "friction" tensor, then one would add terms to the equations of motion of the form

$$\boldsymbol{f}_f = \rho \boldsymbol{w} \cdot \boldsymbol{f} \; . \qquad 2.5.24$$

53

which make the following contribution to the tensor form of Lagrange's identity:

$$\int_V \mathbf{r}\, \boldsymbol{f}_f \, dV = \int_V \mathbf{r}\, \rho\, \mathbf{w} \cdot \mathbf{\mathcal{F}}\, dV = \frac{1}{2} \int_V \rho [\, \frac{d}{dt} (\mathbf{r r}) \,] \cdot \mathbf{\mathcal{F}}\, dV . \qquad 2.5.25$$

If $\mathbf{\mathcal{F}}$ were indeed constant throughout the configuration equation (2.5.25) would just become

$$\frac{1}{2} \left(\frac{d\mathbf{I}}{dt} \right) : \mathbf{\mathcal{F}} \qquad 2.5.26$$

and Lagrange's identity in its full generality would be

$$\frac{1}{2} \frac{d^2 \mathbf{I}}{dt^2} + \frac{1}{2} \left(\frac{d\mathbf{I}}{dt} \right) : \mathbf{\mathcal{F}} + 2\boldsymbol{w} \cdot \mathbf{L} = 2\, [\mathbf{\mathcal{T}} - \mathbf{\mathcal{M}}]$$

$$+ n\mathbf{\mathcal{H}} + \boldsymbol{w} \cdot [\mathbf{I}\boldsymbol{w} - \boldsymbol{w}\mathbf{I}] + \mathbf{I}[(1-\gamma)\mathbf{\mathcal{U}} + \mathbf{\mathcal{M}}] + \mathbf{\mathcal{S}} . \qquad 2.5.27$$

6. Summary

       In this chapter we have continued the development of the virial theorem as it appears in more contemporary usage. The tensor virial theorem is a more general form of Lagrange's identity which when averaged over time provides rather general expressions for the coordinated behavior of some energy like tensors of the system. Further insight into the nature of this process is discovered in the second section where we find that taking higher order spatial moments of the equation of motion is equivalent to recovering the information selectively lost in the classical derivation of the virial theorem. In principle, this approach could be used in a prescription for the complete solution of the equations of motion. However, it seems likely that in practice it would be more difficult than implementing a direct numerical solution of the original equations themselves. The importance of the method lies in the fact that such moment expressions for stable systems are normally rapidly convergent. Thus, the largest amount of information can be recovered with the least effort.

In the next two sections, we consider the effects of a relativity principle on the development of the virial theorem. It is generally observed that large velocities require large gravitational fields to keep them in check and thus one might argue that a separate discussion of the effects of special relativity are not warranted. However, there are at least two dynamically stable systems for which this is not true (i.e., pure radiation spheres which approximate some models of super massive stars and white dwarfs where low-mass, high velocity electrons, were kept in check by the high-mass, low velocity protons). In addition, Lagrange's identity is applicable to systems which are not in equilibrium and hence may be relativistic. For this reason, we have developed the special and general relativistic versions of the theorem separately and will return in the last chapter to discuss some specific applications of them.

I have attempted throughout the chapter to emphasize the similarity of the derivation of the virial theorem and particularly in Section 3, the perceptive reader may have noticed that the derivation is equivalent to carrying out

$$\int_V \mathbf{r} \cdot \Box \cdot \mathbf{J} \, dV = 0 \qquad 2.6.1$$

where $\mathbf{r}$ is a four vector in the Lorentz metric. However, we ignored all contributions from the time-like part. These would have been of the form

$$\frac{1}{2} \frac{d}{dt} \int_V t \, \mathbf{J}_{oo} \, dV = \int_V (t \sum_j \partial \mathbf{J}_{oj}/\partial x_j) dV , \qquad 2.6.2$$

which after appropriate application of the divergence theorem, becomes

$$\frac{d}{dt}\left(\int_V \varepsilon dV\right) = \frac{dE}{dt} = - \int_S \frac{\rho c^2}{\gamma} \frac{d\mathbf{x}}{dt} \cdot d\mathbf{s} . \qquad 2.6.3$$

Because of the linear independence of the time and space coordinates, this is not a new result but rather an expression of the conservation of energy. Essentially it states that the time rate of change of the total energy of the system equals the momentum flux across the system boundary multiplied by c^2.

When the metric is no longer the 'flat' Lorentz metric, as in the case in Section 4, things are no longer simple. It is this loss of simplicity which caused me to stray from the more rigorous approaches of other sections. Thus rather than stress the manipulative complexity of the post-Newtonian approximation, I have attempted to provide physical motivation from the existence of terms arising in the equations of motion resulting from the non-linear nature of general relativity. The derivation presented in this section follows exactly the prescription of earlier sections, but for simplicity I presented the development of the scalar version of the virial theorem only. Sticking with the post-Newtonian approximation avoided some difficult problems of uniqueness and interpretation. However, I remain convinced that a general formulation of Lagrange's identity and the virial theorem which is compatible with the field equation of Einstein exists and its formulation would be most rewarding. For most stellar astrophysical applications, the post-Newtonian result is probably sufficient.

In the last section of this chapter some of the powers of the virial theorem to deal with difficult situations became apparent. The results which have been generalized to include additional effects are not the result of any new physical concepts. Rather, they are the result of the specific identification of the physical contribution to the system made by such attributes as magnetic fields and macroscopic motion. Although I included only magnetic fields throughout most of the discussion, the inclusion of electric fields in the Maxwell field tensor makes it clear how to proceed should they be present. Lastly we looked again at velocity dependent forces not so much with an eye to their effect on the virial theorem, but rather with a view to their persistent presence in the variational form of Lagrange's identity.

The presence of all these complicating aspects is included only to make their interplay explicit. The basic theorem must hold, all the rest is done to glean more insight.

Notes to Chapter 2

2.1 The term on the left of 2.1.2 simply becomes

$$\int_V \rho \mathbf{r} \frac{d\mathbf{u}}{dt} dV = \int_V \rho \mathbf{r} \frac{d^2\mathbf{r}}{dt^2} dV = \int_V \rho \frac{d}{dt}\left(\frac{\mathbf{r}\,d\mathbf{r}}{dt}\right)dV - \int_V \rho \frac{d\mathbf{r}}{dt}\frac{d\mathbf{r}}{dt} dV \ . \qquad \text{N2.1.1}$$

The third term can further be simplified so that

$$\int_V \rho \frac{d}{dt}\left(\mathbf{r}\frac{d\mathbf{r}}{dt}\right)dV = \frac{1}{2} \int_V \rho \frac{d^2}{dt^2}(\mathbf{r}\mathbf{r})dV = \frac{1}{2}\frac{d^2}{dt^2}\int_V \rho \mathbf{r}\mathbf{r}\, dV \ . \qquad \text{N2.1.2}$$

In obtaining the third term in equation 2.1.4 we have assumed that the volume V is large enough to contain all matter and the conservation of mass argument explicitly developed in Chapter 1, Section 4, Equation 1.4.6. Using the results of N2.1.1 and N2.1.2 we get

$$\frac{1}{2}\frac{d^2}{dt^2}\int_V (\rho\mathbf{r}\mathbf{r})dV = \int_V (\rho\mathbf{u}\mathbf{u})dV + \int_V \rho \mathbf{r} \nabla\Phi dV \ . \qquad \text{N2.1.3}$$

2.2 If one writes equation 2.1.9 in component form, one gets

$$\frac{1}{2}\frac{d^2 \mathfrak{I}_{ij}}{dt^2} = 2\mathfrak{T}_{ij} + n\,\mathfrak{U}_{ij} \qquad \text{N2.2.1}$$

where in cartesian coordinates the tensor components are

$$\mathfrak{I}_{ij} = \int_V \rho\, x_i x_j\, dV$$

$$\mathfrak{T}_{ij} = \int_V \rho\, u_i u_j\, dV \qquad \text{N2.2.2}$$

$$\mathfrak{U}_{ij} = \frac{1}{2}\int_V\int_{V'} \rho(\mathbf{r})\rho(\mathbf{r}')\,(x_i-x_i')(x_j-x_j')(|\mathbf{r}-\mathbf{r}'|)^{n-2}\,dV' dV \ .$$

By contracting these tensors and letting the potential be that of the gravitational field, we recover the scalar form of Lagrange's identity as given in Chapter 1, Equation 1.4.12.

2.3 Taking the scalar product of a space-like position vector with 2.3.6 and integrating over a volume sufficient to contain the system we get

$$c\int_V \sum_j x_j \frac{\partial(\rho u_j)}{\partial t} dV + \sum_j \sum_k \int_V x_j \frac{\partial \mathfrak{I}_{jk}}{\partial x_k} dV = 0 \ . \qquad \text{N2.3.1}$$

From the chain rule the second integral can be written as

$$\sum_j \sum_k \int_V x_j \frac{\partial \mathfrak{I}_{jk}}{\partial x_k} dV = \sum_j \sum_k \int_V \frac{\partial}{\partial x_k}(x_j \mathfrak{I}_{jk})\, dV - \sum_j \sum_k \int_V \frac{\partial x_j}{\partial x_k} \mathfrak{I}_{jk}\, dV \ . \qquad \text{N2.3.2}$$

The first integral on the right is just the integral of the divergence of $x_j \mathfrak{I}_{jk}$ over V and if the volume is so chosen to enclose the entire system the integral must vanish as $\mathfrak{I}_{jk} = 0$ on the surface enclosing V (i.e., Gauss's Law applies). Since $\partial x_j/\partial x_k = \delta_{jk}$ the second integral just becomes $\sum_j \int_V \mathfrak{I}_{jj} dV$. Equation N2.3.1 is then

$$\sum_j c\int_V x_j \frac{\partial}{\partial t}(\rho u_j)dV - \sum_j \int_V \mathfrak{I}_{jj}\, dV = 0 \ . \qquad \text{N2.3.3}$$

Now, $\partial x_j/\partial t = 0$ from the orthonogonality of the Lorentz frame and $\sum_{j=1}^{3} \mathfrak{I}_{jj} = \sum_{\alpha=1}^{3} \mathfrak{I}_{\alpha\alpha} - \mathfrak{I}_{oo}$ so we can write

$$c \int_V \frac{\partial}{\partial t} (\Sigma_j \rho x_j u_j) \, dV + \int_V \rho c^2 \gamma dV + \int_V \mathfrak{z}_{oo} \, dV = 0 \; . \qquad \text{N2.3.4}$$

With the sign convection $\mathfrak{z}_{oo}$ is the negative of the total energy density. Bergmann[17], among others, shows us that the "relativistic" kinetic energy density τ is given by

$$\tau = \rho c^2 (\gamma^{-1} - 1)$$

or $\qquad \gamma \tau = \rho c^2 - \rho c^2 \gamma \qquad$ N2.3.5

Thus, using 2.3.3 to re-write the first term of N2.3.4, we get

$$\sum_{j=1} \int_V \frac{\partial}{\partial t} [\frac{1}{2} \rho \frac{d(x_j x_j / \gamma)}{dt}] \, dV + \int_V (\rho c^2 - \gamma \tau - \varepsilon - \rho c^2) dV = 0 \; . \qquad \text{N2.3.6}$$

where ε is the potential energy density. Applying Leibnitz's law for the differentiation of definite integrals[18] to the first term in N2.3.6 and re-writing the second one, we get

$$\sum_{j=1} \frac{d}{dt} \int_V \frac{\rho}{2} \frac{d}{dt} (x_j x_j / \gamma) dV = \Omega + T + \int_V \gamma \tau \, dV. \qquad \text{N2.3.7}$$

2.4
$$\mathfrak{z} = \frac{1}{4\pi} [\mathbf{DD} + c^2 \mathbf{HH} - \frac{1}{2} \mathbf{I} (D^2 + c^2 H^2)] \; . \qquad \text{N2.4.1}$$

If we take the divergence of $\mathfrak{z}$ we get

$$\nabla \cdot \mathfrak{z} = \frac{1}{4\pi} \{ (\mathbf{D} \cdot \nabla) \mathbf{D} + \mathbf{D} \nabla \cdot \mathbf{D} + (c^2 \mathbf{H} \cdot \nabla) \mathbf{H} + c^2 \mathbf{H} (\nabla \cdot \mathbf{H}) - \frac{1}{2} \nabla (\mathbf{D} \cdot \mathbf{D} + c^2 \mathbf{H} \cdot \mathbf{H}) \; , \qquad \text{N2.4.2}$$

and invoking Maxwell's laws that $\nabla \cdot \mathbf{D} = \rho_e$ and $\nabla \cdot \mathbf{H} = 0$, this becomes

$$\nabla \cdot \mathfrak{z} = \frac{1}{4\pi} \{ \mathbf{D}\rho + (\mathbf{D} \cdot \nabla) \mathbf{D} + (c^2 \mathbf{H} \cdot \nabla) \mathbf{H} - \frac{1}{2} \nabla (\mathbf{D} \cdot \mathbf{D}) + \frac{1}{2} c^2 \nabla (\mathbf{H} \cdot \mathbf{H}) \} \; . \qquad \text{N2.4.3}$$

Making use of the vector identity

$$\nabla (\mathbf{A} \cdot \mathbf{G}) = \mathbf{A} \times (\nabla \times \mathbf{G}) + (\mathbf{A} \cdot \nabla) \mathbf{G} + \mathbf{G} \times (\nabla \times \mathbf{A}) + (\mathbf{G} \cdot \nabla) \mathbf{A} \qquad \text{N2.4.4}$$

equation N2.4.3 takes on the more familiar form

$$\nabla \cdot \mathfrak{z} = \frac{1}{4\pi} \{ \mathbf{D} \rho_e - \mathbf{D} \times (\nabla \times \mathbf{D}) + c^2 [\mathbf{H} \times (\nabla \times \mathbf{H})] \} \qquad \text{N2.4.5}$$

2.5 By use of the vector identity (N2.4.4), it is clear that

$$\frac{1}{2} \nabla (\mathbf{\omega} \times \mathbf{r})^2 = (\mathbf{\omega} \times \mathbf{r}) \times [\nabla \times (\mathbf{\omega} \times \mathbf{r})] + [(\mathbf{\omega} \times \mathbf{r}) \cdot \nabla] (\mathbf{\omega} \times \mathbf{r}) \; , \qquad \text{N2.5.1}$$

but

$$\nabla \times (\mathbf{\omega} \times \mathbf{r}) = (\mathbf{\omega} \cdot \nabla) \mathbf{r} - (\mathbf{r} \cdot \nabla) \mathbf{\omega} - \mathbf{\omega} (\nabla \cdot \mathbf{r}) + \mathbf{r} (\nabla \cdot \mathbf{\omega}) = 0 \; . \qquad \text{N2.5.2}$$

The constancy of $\mathbf{\omega}$ causes the second and fourth term to vanish while the first and third terms are equal but of opposite sign. Again, since $\mathbf{\omega}$ = const

$$[(\mathbf{\omega} \times \mathbf{r}) \cdot \nabla] (\mathbf{\omega} \times \mathbf{r}) = \mathbf{\omega} \times [(\mathbf{\omega} \times \mathbf{r}) \cdot \nabla](\mathbf{r}). \qquad \text{N2.5.3}$$

In component form

$$[(\mathbf{\omega} \times \mathbf{r}) \cdot \nabla] (\mathbf{r}) = \sum_i \sum_j \sum_k \epsilon_{ijk} \omega_j r_k \frac{\partial x_\ell}{\partial x_i} = \sum_j \sum_k \epsilon_{\ell jk} \omega_j r_k = \mathbf{\omega} \times \mathbf{r} \qquad \text{N2.5.4}$$

where ϵ_{ijk} in a completely antisymmetric tensor of rank 3 sometimes called the Levi Civita tensor density. Thus

$$\frac{1}{2} \nabla (\mathbf{\omega} \times \mathbf{r})^2 = \mathbf{\omega} \times (\mathbf{\omega} \times \mathbf{r}) \qquad \text{N2.5.5}$$

2.6

$$\int_V \rho \mathbf{r} \frac{d\mathbf{w}}{dt} dV + 2 \int_V \rho \mathbf{r}(\pmb{w}\times\mathbf{w}) dV = -\int_V \rho \nabla[\Phi + (\pmb{w}\times\mathbf{r})^2 + P/\rho] dV - \frac{1}{4\pi} \int_V \mathbf{r}[\mathbf{H}\times(\nabla\times\mathbf{H})] dV.$$ N2.6.1

As in Section 3, the first term on the left becomes

$$\int_V \rho \mathbf{r} \frac{d\mathbf{w}}{dt} dV = \frac{1}{2} \frac{d^2}{dt^2}\left(\int_V \rho(\mathbf{r}\,\mathbf{r}) dV\right) - \int_V \rho(\mathbf{w}\,\mathbf{w}) dV \ .$$ N2.6.2

The second term is more difficult to simplify. Let the local velocity field $\mathbf{w} = \widetilde{\pmb{w}} \times \mathbf{r}$ by defining a local angular velocity field $\widetilde{\pmb{w}}$. Then, by expanding the resulting vector triple product in the second term we can write

$$2\int_V \rho \mathbf{r}(\pmb{w}\times\mathbf{w}) dV = 2\int_V \rho \mathbf{r}[\pmb{w}\times(\widetilde{\pmb{w}}\times\mathbf{r})] dV = 2\int_V \rho \mathbf{r}[\widetilde{\pmb{w}}(\pmb{w}\cdot\mathbf{r}) - \mathbf{r}(\pmb{w}\cdot\widetilde{\pmb{w}})] dV$$

or $\quad 2\int_V \rho \mathbf{r}(\pmb{w}\times\mathbf{w}) dV = 2\pmb{w}\cdot\int_V \rho[\mathbf{r}(\mathbf{r}\,\widetilde{\pmb{w}}) - \widetilde{\pmb{w}}(\mathbf{r}\,\mathbf{r})] dV = 2\pmb{w}\cdot\mathbf{\mathcal{L}}$ N2.6.3

where the tensor $\mathbf{\mathcal{L}}$ is an angular moment-like three index tensor representing the various components of volume net angular momentum within the body. Thus, $\pmb{w}\cdot\mathbf{\mathcal{L}}$ is a kinetic energy like tensor resulting from the net motions induced by the coriolis forces. However, since there can be no net motions about any axis other than that defined by the total angular momentum of the body all components of $\pmb{w}\cdot\mathbf{\mathcal{L}}$ must be zero except those associated with the axis of rotation. In addition we can choose our rotating frame $\pmb{w}$ so that in that frame the net angular momentum is zero and all contributions from the term $\pmb{w}\cdot\mathbf{\mathcal{L}}$ will vanish. For the sake of generality we shall keep the term for the present.

The first term on the right of equation N2.6.1 is given by equation 2.1.4. Since we just went to some length to show that $\pmb{w}\times(\pmb{w}\times\mathbf{r}) = \frac{1}{2}\nabla(\pmb{w}\times\mathbf{r})^2$, we shall use the earlier version and an expansion of the vector triple product to evaluate the second term on the right. Thus

$$\frac{1}{2}\int_V \rho \mathbf{r}\nabla(\pmb{w}\times\mathbf{r})^2 dV = \int_V \rho \mathbf{r}[\pmb{w}\times(\pmb{w}\times\mathbf{r})] dV = \int_V \rho \mathbf{r}\pmb{w}(\pmb{w}\cdot\mathbf{r}) dV - \int_V \rho \mathbf{r}\mathbf{r}\,\omega^2 dV$$

or $\quad \int_V \rho \mathbf{r}\nabla(\pmb{w}\times\mathbf{r})^2 dV = \pmb{w}\cdot\left(\int_V \rho \mathbf{r}\mathbf{r}\,\pmb{w} dV\right) - \omega^2\int_V \rho \mathbf{r}\mathbf{r}\,dV \ .$ N2.6.4

It is worth noting at this point that all terms dealt with so far involve only the volume integrals $\int_V \rho \mathbf{r}\mathbf{r}\,dV$ and $\int_V \rho \mathbf{w}\mathbf{w}\,dV$ which are the same as the tensors defined in Section 1. Thus, by combining N2.6.2, N2.6.3, N2.6.4, and 2.1.4, we may assess our progress in simplifying (N2.6.1) so far (see 2.5.10).

References

1. Rayleigh, L. (1903), Scientific Papers, Vol. 4, Cambridge, England, p. 491.
2. Parker, E. N. (1954), Phys. Rev. 96, p. 1686-9.
3. _____. (1957), Ap. J. Suppl. 3, p. 51.
4. Chandrasekhar, S., and Fermi, E. (1953), Ap. J. 118, p. 116.
5. Lebovitz, N. R. (1961), Ap. J. 634, p. 500-536.
6. Chandrasekhar, S., and Lebovitz, N. R. (1962), Ap. J. 135, p. 238, 248, 1032.
7. _____. (1962), Ap. J. 136, p. 1037-1047.
8. Chandrasekhar, S. (1961), Hydrodynamic and Hydromagnetic Stability, Oxford University Press, London, p. 571-581.
9. _____. (1969), Ellipsoidal Figures of Equilibrium, Yale University Press, Chapt. 2, p. 15-37.
10. Chandrasekhar, S., and Lebovitz, N. R. (1962), Ap. J. 135, p. 238-247.
11. _____. (1962), Ap. J. 136, p. 1032-1036.
12. Chandrasekhar, S. (1962), Ap. J. 136, p. 1069-1081.
13. Chandrasekhar, S., and Lebovitz, N. R. (1968), Ap. J. 152, p. 293-304.
14. Chandrasekhar, S. (1957), An Introduction to the Study of Stellar Structure, Dover Pub., Inc., p. 424, eq. 81.
15. Landau, L. D. and Lifshitz, E. M. (1962), The Classical Theory of Fields, Trans. M. Hamermesh, Addison-Wesley Pub. Co., Reading, Mass. USA, p. 95-97.
16. Misner, C. W., Thorne, K. S., and Wheeler, J. A. (1973), Gravitation, W. H. Freeman & Co., San Francisco.
17. Bergmann, P. G. (1942), Introduction to the Theory of Relativity, Prentice-Hall, Inc., New York, p. 92.
18. Sokolnikoff, I. S., and Redheffer, .(1958), Mathematics of Physics and Moderning Engineering, McGraw-Hill Book Co., Inc., New York, p. 262.
19. Einstein, A., Infeld, E., and Hoffman, (1938), Ann. Math 39, p. 65-100.
20. Chandrasekhar, S. (1965), Ap. J. 142, p. 1488-1512.
21. Schiff, I. I. (1960), American Journal of Physics 28, 340.
22. Misner, C. W., Thorne, K. S., and Wheeler, J. A. (1973), Gravitation, W. H. Freeman & Co., San Francisco, p. 1082.
23. Chandrasekhar, S. (1965), Ap. J. 142, p. 1499, eq. 80.
24. Chandrasekhar, S., and Nutku, Y. (1969), Ap. J. 158, p. 55-79.
25. Chandrasekhar, S., and Esposita, P. (1970), Ap. J. 160, p. 153-179.
26. Ni, W. (1973), Ap. J. 181, p. 957-976.
27. Panofsky, W. K. H. and Phillips, M. (1955), Classical Electricity and Magnetism, Addison-Wesley Pub. Co., Reading, Mass. p. 162.
28. Goldstein, H. (1959), Classical Mechanics, Addison-Wesley Pub. Co., Reading, Mass. p. 133.
29. Milne, E. A. (1925), Phil. Mag. S. 6, Vol. 50, p. 409-414.

3. The Variational Form of the Virial Theorem

1. Variations and Perturbations and their Implications for the Virial Theorem

    Perturbation analysis is truly an old mechanism by which one explores the behavior of a system in a known state by assuming there are small variations in the independent variables describing the system and determining the individual variation in the independent variables. The vehicle for determining the independent variable changes, is found in the very equations which describe the initial state of the system. The equations usually chosen for this type of analysis are the equations of motion for the system. For example, consider the equations of motion for an object moving under the influence of a point potential Ω.

$$\frac{d^2 \mathbf{r}}{dt^2} = - \nabla \Phi \qquad \qquad 3.1.1$$

    Assuming a solution $\mathbf{r}_o(t)$ is known, which satisfies 3.1.1 for a particular potential Φ_o. Since 3.1.1

is valid for any system where Φ is known, one could define

$$\left.\begin{array}{r}\Phi = \Phi_o + \delta\Phi \\ \mathbf{r} = \mathbf{r}_o + \delta\mathbf{r}\end{array}\right\}, \qquad 3.1.2$$

and 3.1.1 would require

$$\frac{d^2}{dt^2}(\mathbf{r}_o + \delta\mathbf{r}) = -\nabla[\Phi_o + \delta\Phi] \ . \qquad 3.1.3$$

However, since both ∇ and d/dt are linear operators, 3.1.3 becomes

$$\frac{d^2\mathbf{r}_o}{dt^2} + \frac{d^2\delta\mathbf{r}}{dt^2} = -\nabla\Phi_o - \nabla(\delta\Phi) \ , \qquad 3.1.4$$

but we already know that

$$\frac{d^2\mathbf{r}_o}{dt^2} = -\nabla\Phi_o \qquad 3.1.5$$

so that by subtracting 3.1.5 from 3.1.4, we get

$$\frac{d^2\delta\mathbf{r}}{dt^2} = -\nabla(\delta\Phi) \qquad 3.1.6$$

which we called the perturbed equations of motions where $\delta\Phi$ is the perturbation that involves the perturbation $\delta\mathbf{r}$. A short approach which leads to the same result is to "take the variation" of equation 3.1.1 wherein the operator δ is not affected by time or space derivatives. This technique "works" because the time and space operators in the equation of motion are linear, hence any linear perturbation or departure from a given solution will produce the sum of the original equations of motion on the perturbed equations of motion. In general, I shall use the variational operator δ in this sense, that is, it represents a small departure of a variable from the value it had which satisfies the equations governing the system.

It is not necessary that one perturbs the equations of motion in order to gain information about the system. Clearly any equations which describe the structure of the system are subject to this type of analysis. Thus, if taking variations of the equations of motion produces useful results, might not the variational form of the moments of those equations also be expected to contain interesting information? It was in this spirit that Paul Ledoux developed the variational form of the scalar virial theorem[1], and was able to predict the global behavior of a star.

The variational approach yields differential equations which describe parameter relationships for a system disturbed from an initial state. If that state happens to be an equilibrium state, variational analysis of the equations of motion would yield a description of the system motion about the equilibrium configuration. Variational analysis of spatial moments could then be expected to yield macroscopic properties of that motion. This is indeed the case, as Ledoux demonstrated by determining perhaps the most obvious macroscopic property of such motion, the pulsational period of the system. Chandrasekhar[2] found the tensor form of the virial theorem useful in determining non-radial modes of oscillation of stars. In addition he and Fermi[3] investigated the effects of a magnetic field on the pulsation of a star. An additional macroscopic property closely connected with the pulsational period with which this approach deals is the global stability of the system. We shall examine this aspect of the analysis later. For now, let us be content with observing in some detail how the variational approach yields the pulsational periods of stars.

2. Radial Pulsations for Self-Gravitating Systems

In this section we shall use the virial theorem, to obtain an expression for the frequency of radial pulsations in a gas sphere. The approach will be to apply a small variation to the virial theorem and by making use of several conservation laws, obtain expressions for the variation of the moment of inertia, kinetic energy, and potential energy as a function of time.

Remember from the earlier section that Lagrange's identity is

$$\frac{1}{2}\frac{d^2 I}{dt^2} = 2T + \Omega \ . \qquad 3.2.1$$

In this form, no time averaging has been carried out and the equation must apply to a dynamic system at any point in time. Now, consider a star with radius R. Let r be the distance from the center of symmetry to any point in the configuration and δr be the displacement of a point mass from the equilibrium position r_o. Conservation of mass requires that for a spherical shell of radius r

$$m(r_o + \delta r) = m(r_o) \ . \qquad 3.2.2$$

We wish to find the variations δI, δT, $\delta \Omega$ of the quantities I, T, Ω, from the equilibrium values I_o, T_o, and Ω_o. The variational form of the virial theorem then becomes

$$\frac{1}{2} \frac{d^2(\delta I)}{dt^2} = 2\delta T + \delta \Omega \ . \qquad 3.2.3$$

Since I was defined as the moment of inertia about the center of the coordinate system, we have by definition that,

$$I = \int_o^M r^2 dm(r) \qquad 3.2.4$$

Thus, we have

$$\delta I = \int_o^M 2r\delta r\, dm(r) + \int_o^M r^2 \delta(dm(r)) \qquad 3.2.5$$

Since, by the conservation of mass (equation 3.2.2), $\delta m(r) = 0$ for all r, $d(\delta m(r)) = \delta(dm(r)) = 0$ for all r and the second integral of (3.2.5) vanishes leaving

$$\delta I = \int_o^M 2r\delta r\, dm(r) \ . \qquad 3.2.6$$

Now from equation (3.2.4), we have

$$\frac{dI}{dm(r)} = r^2 \ . \qquad 3.2.7$$

Since this must always be true, it is true at the equilibrium point r_o. Therefore,

$$dI_o = r_o^2 dm(r) \qquad 3.2.8$$

So, to first order accuracy in r, we may re-write (3.2.6) as

$$\delta I = 2 \int_0^{I_o} \frac{\delta r}{r_o} \, dI_o \qquad 3.2.9$$

In a similar manner we may evaluate the variation of the gravitational potential energy with respect to small variation in $r^{3.1}$ and obtain

$$\delta \Omega = - \int_0^M \frac{\delta r}{r_o} \, d\Omega \quad . \qquad 3.2.10$$

All that now remains to be determined in equation (3.2.3) is the variation of the total kinetic energy T. To first order only the variation of the thermal kinetic energy will contribute to 3.2.3.[3.2]

$$2\delta T \simeq 3 \int_0^M \frac{P_o}{\rho_o} (\gamma-1) \frac{\delta\rho}{\rho_o} \, dm(r) \qquad 3.2.11$$

In order to facilitate obtaining an expression for $\delta\rho/\rho_o$, we shall now specify a time dependence for the pulsation about r_o. For simplicity, let us assume the motion is simply periodic. Thus, defining a quantity ξ as

$$\xi = \frac{\delta r}{r_o} = \xi_o \, e^{i\sigma t} \qquad 3.2.12$$

where $2\pi/\sigma$ is the period of oscillation, we may re-write the variations of I and Ω as follows:

$$\delta I = 2e^{i\sigma t} \int_0^M \xi_o \, dI_o$$

$$\delta \Omega = - e^{i\sigma t} \int_0^M \xi_o \, d\Omega_o \quad . \qquad 3.2.13$$

Conservation of mass requires that[3.3]

$$\frac{\delta\rho}{\rho_o} = - (3\xi_o + r_o \frac{d\xi_o}{dr_o}) \, e^{i\sigma t} . \qquad 3.2.14$$

Substitution of this back into the expression for the variation of the kinetic energy yields

$$2\delta T = - 3 \int_0^M \frac{P_o}{\rho_o} (\gamma-1) [3\xi_o + r_o \frac{d\xi_o}{dr_o}] \, e^{i\sigma t} \, dm(r_o) . \qquad 3.2.15$$

Equation 3.2.15 may be simplified to yield[3.4]

$$2\delta T = 3\, e^{i\sigma t} \int_0^M \frac{P_0 \xi_0}{\rho_0} \frac{d}{dr} dm(r) + 3\, e^{i\sigma t} \int_0^{\Omega_0} \xi_0 (\gamma-1) d\Omega_0 \; . \qquad 3.2.16$$

We now have all the material necessary to evaluate the variational form of the virial theorem to first order accuracy. Substituting (3.2.13) and (3.2.16) into (3.2.3), we obtain

$$-\sigma^2 e^{i\sigma t} \int_0^{I_0} \xi_0 dI_0 = 3 e^{i\sigma t} \int_0^M \frac{P_0 \xi_0 r_0}{\rho_0} \frac{d\gamma}{dr} dm(r) \qquad 3.2.17$$
$$+ 3\, e^{i\sigma t} \int_0^{\Omega_0} \xi_0(\gamma-1) d\Omega_0 - e^{i\sigma t} \int_0^{\Omega_0} \xi d\Omega_0 .$$

Solving for σ^2, which is related to the pulsation period, we have

$$\sigma^2 = \frac{-\int_0^{\Omega_0} (3\gamma-4)\xi_0 d\Omega_0 + 3 \int_0^M \frac{P_0 \xi_0 r_0}{\rho_0} \frac{d\gamma}{dr} dm(r)}{\int_0^{I_0} \xi_0 dI_0} \qquad 3.2.18$$

For a model of known equilibrium structure, the integrals in (3.2.18) may be evaluated and the frequency for which it is stable to radial pulsations may be computed. However, for purposes of examining the behavior of a pulsating star we may assume the star is homogeneous so that γ is constant. Also, let us assume the pulsation increases radially outward in a linear manner. Under these admittedly ad hoc assumptions, equation (3.2.18) reduces to the extremely simple form

$$\sigma^2 = -\frac{(3\gamma-4)\Omega_0}{I_0} \; . \qquad 3.2.19$$

In order to obtain a feeling for the formula we have developed, we shall attempt to estimate some approximation pulsation frequencies. For a sphere of uniform density

$$\Omega_0 = \frac{3}{5} \frac{GM^2}{R_0} \; . \qquad 3.2.20$$

The moment of inertia for a sphere about an axis is equal to 3/2 the moment of inertia about its center and is given by

$$I_z = \frac{2}{5} MR_o^2 = \frac{3}{2} I_o \; .\qquad 3.2.21$$

Therefore,

$$I_o = \frac{4\,MR_o^2}{15} \; .\qquad 3.2.22$$

Prior theory concerning stellar structure implies that $\gamma > 4/3$. If we take $\gamma = 5/3$, we obtain

$$\sigma^2 = \frac{9}{4} \frac{GM}{R_o^3} \; ,\qquad 3.2.23$$

or

$$\sigma^2 = 3\pi G \bar{\rho} \; .$$

Remembering that the period T is just $2\pi/\sigma$, we have

$$T = \left(\sqrt{\frac{4\pi}{3\,G}}\right) \rho^{-\frac{1}{2}} \; ,\qquad 3.2.24$$

Thus, we see that the theory does produce a period which is inversely proportional to the square root of the mean density. This law has been found to be experimentally correct in the case of the classical cepheids. It should be noted that this property will be preserved even for the integral form (3.2.18), only the constant of proportionality will change. If we evaluate the constant of proportionality from equation (3.2.24), we have

$$T \simeq 7.95 \times 10^{-3} \rho^{-\frac{1}{2}} \quad \text{sec.} \qquad 3.2.25$$

where $\bar{\rho}$ is given in gm/cc.

Taking an observed value for the mean density of a cepheid variable to be between 10^{-3} and 10^{-6} cm/cc (Ledeoux and Walraven)[4], we arrive at the following estimate for the periods of these stars.

$$0.3 \text{ days} < T < 90 \text{ days} \qquad 3.2.26$$

It is freely admitted that this estimate is arrived at in the crudest way, however, and it is comforting that the result nicely brackets the observed periods for cepheid variables. It should also be noted that for most stars the expression arrived at in (3.2.23) for σ^2 is a lower limit. As the mass becomes more centrally concentrated the magnitude of the gravitational energy will increase while the moment of inertia will decrease. Even for reasonable density distributions the value arrived at in (3.2.23) will not differ by more than an order of magnitude. This would imply that a value for the period calculated in this manner should be correct within a factor of 2 or 3. Thus, without solving the force equations outlined, an estimate for a very important parameter in describing the pulsation of a gas sphere may be obtained which is the period for which that sphere is stable to radial pulsation.

3. The Influence of Magnetic and Rotational Energy Upon a Pulsating System

We shall now consider what the effect of introducing magnetic and rotational energies into a pulsating system will be upon the frequency of pulsation of that system. It is worth noting that solution of such a problem in terms of the force equations would be difficult indeed as it would require detailed knowledge of the geometry of the magnetic field throughout the star. However, since our approach expresses the pulsation frequency in terms of volume integrals, only knowledge of the total magnetic and rotational energies will be required.

In order to simplify the mathematical development we shall make some of the assumptions which were made during previous sections. These assumptions are listed below:

1) A first order theory will be adequate. That is, all deviations from equilibrium shall be small.
2) Radiation pressure will be considered to be negligible (i.e., $\Gamma_1 = \gamma$).
3) γ will be constant throughout the system.

We have already seen that it is possible to write Lagrange's identity so as to include the effects of rotational and magnetic energy. One of the points of that derivation that required some care was the inclusion of surface terms arising from the fact that a stellar magnetic field usually extends well beyond the normal surface of the star. However, for the moment let us neglect these terms since they usually will be small and as such will not affect the general character of the solution. Thus, as in Chapter 2, we may write the scalar form of Lagrange's identity as follows

$$\frac{1}{2}\frac{d^2 I}{dt^2} = 2T + \Omega + \mathfrak{M} \qquad 3.3.1$$

where T is the total kinetic energy including rotation and $\mathfrak{M}$ is the total magnetic energy.

Now let us break up the total kinetic energy of the system into the sum of three energies $\mathfrak{T}_1$, $\mathfrak{T}_2$ and $\mathfrak{T}_3$ where $\mathfrak{T}_1$ is the kinetic energy of the pulsating system due to the pulsating motion, $\mathfrak{T}_2$ is the kinetic energy of the gas due to thermal energy, and $\mathfrak{T}_3$ is the kinetic energy of rotation.

The contribution to the kinetic energy (of particle motion) due to an element of mass is

$$d\mathfrak{T}_2 = 3/2 \; kTdN = 3/2 \; RTdm = 3/2(c_p - c_v)Tdm \; . \qquad 3.3.2$$

But the internal energy $d\mathfrak{U}$ of the element of mass is

$$d\mathfrak{U} = c_v Tdm \; , \qquad 3.3.3$$

Combining (3.3.2) and (3.3.3), with the definition of γ, we have

$$d\mathfrak{T}_2 = 3/2 \; (\gamma - 1)d\mathfrak{U} \; . \qquad 3.3.4$$

Integrating this over the entire system we obtain

$$2\mathfrak{T}_2 = 3 \; (\gamma - 1)\mathfrak{U} \; . \qquad 3.3.5$$

Thus, we may write the virial theorem for the system as

$$\frac{1}{2}\frac{d^2 I}{dt^2} = 2\mathfrak{T}_1 + 2\mathfrak{T}_3 + 3(\gamma - 1)\mathfrak{U} + \Omega + \mathfrak{M} \; . \qquad 3.3.6$$

The variational form becomes

$$\frac{1}{2}\frac{d^2}{dt^2}(\delta I) = 2\delta\mathcal{T}_3 + 3(\gamma-1)\delta\mathcal{U} + \delta\mathfrak{M} + \delta\Omega \qquad 3.3.7$$

There is no term containing a variation in γ as it is assumed to be constant throughout the system. In Section 2, it was shown that variation of the pulsational kinetic energy was of second order and could therefore be neglected. Thus

$$2\delta\mathcal{T}_1 = 0 \ . \qquad 3.3.8$$

Since we have already computed the variation of the kinetic energy of the gas, we may most easily find the variation of the total internal energy ($\delta\mathcal{U}$) in terms of this quantity. As a result of the assumption of constant γ, we may take the variation of equation (3.3.5), and obtain

$$2\delta\mathcal{T}_2 = 3(\gamma - 1)\delta\mathcal{U} \ . \qquad 3.3.9$$

Now, if we further assume a periodic form for the pulsation and a linearly increasing amplitude (ξ_o = const.), equation (3.2.15), gives the following expression for the variation of the kinetic energy of the gas:

$$2\delta\mathcal{T}_2 = -3 \int_o^M \frac{3P_o\xi_o}{\rho_o}(\gamma - 1)e^{i\sigma t} dm(r_o) \qquad 3.3.10$$

or $\quad 2\delta\mathcal{T}_2 = -3(\gamma - 1)\xi_o e^{i\sigma t} \int_V 3P_o dV$

Combining (3.3.10) and (3.3.9), we obtain

$$\delta\mathcal{U} = -\xi_o e^{i\sigma t} \int_V 3P_o dV_o \ . \qquad 3.3.11$$

In order to express this variation in terms of the other energies present in the system, we shall assume that the system is in quasi-steady state. With this assumption, the relevant quantities are averaged over one pulsation period so that the $\langle \frac{d^2 I}{dt^2} \rangle = 0$. We shall assume that the remaining values are the equilibrium values of the configuration. The virial theorem as expressed in (3.3.6) then becomes

$$2\mathcal{T}_1(0) + 2\mathcal{T}_2(0) + 2\mathcal{T}_3(0) + \Omega_o + \mathfrak{M}_o = 0 \qquad 3.3.12$$

Now, from the elementary kinetic theory of gases we have

$$\mathcal{T}_2 = 3/2 \int_V P_o dV \qquad 3.3.13$$

Also, since the system is neither expanding or contracting,

$$\mathcal{T}_1(0) = <\mathcal{T}_1> = 0 \qquad 3.3.14$$

Making use of these two results and replacing the average values of the quantities in (3.3.12) with the equilibrium values we find

$$-3 \int_V P_o dV = 2\mathcal{T}_3(0) + \Omega_o + \mathfrak{M}_o \qquad 3.3.15$$

Identifying the left side of (3.3.15) with the right side of (3.3.11), we finally obtain

$$\delta\mathcal{U} = \xi_o e^{i\sigma t} (2\mathcal{T}_3(0) + \Omega_o + \mathfrak{M}_o) . \qquad 3.3.16$$

We have already determined expressions for the variations of the gravitational energy and moment of inertia in the previous section (equation 3.2.13). Under the assumption used above, of a constant perturbation ξ_0, these equations become

$$\delta I = 2e^{i\sigma t} \xi_o I_o \qquad 3.3.17$$

and $$\delta\Omega = -e^{i\sigma t} \xi_o \Omega_o$$

Thus, we need only obtain expressions for the variation of the magnetic and rotational energies in order to evaluate equations (3.3.7).

Consider first the rotational energy. Now an element of mass rotating about an axis with a velocity ω will possess an elemental angular momentum

$$d\mathcal{L} = \omega(x^2 + y^2)dm(r) = \omega r^2 \sin^2\theta \, dm(r) \qquad 3.3.18$$

71

Here the x-y plane is the plane perpendicular to axis of rotation and θ is the polar angle measured from the axis of rotation. Such an elemental mass possessing such an angular momentum will have a rotational kinetic energy given by

$$d\mathcal{T}_3 = 1/2\ \omega d\mathcal{L}\ . \qquad 3.3.19$$

Thus, the total rotational kinetic energy is just

$$\mathcal{T}_3 = \int_0^{\mathcal{L}} 1/2\ \omega d\mathcal{L}\ . \qquad 3.3.20$$

We may use this expression to obtain the variation $\delta\mathcal{T}_3$. The first term on the right of equation (3.3.7) becomes

$$2\delta\mathcal{T}_3 = \delta\int_0^{\mathcal{L}}\omega d\mathcal{L} = \int_0^{\mathcal{L}}\delta\omega d\mathcal{L} + \int_0^{\mathcal{L}}\omega d(\delta\mathcal{L})\ . \qquad 3.3.21$$

However, the conservation of angular momentum requires that $\mathcal{L}$ remain constant during the pulsation. Thus, the variation of $\mathcal{L}$ is zero and the last integral on the right of (3.3.21) vanishes, so that

$$2\delta\mathcal{T}_3 = \int_0^{\mathcal{L}}\delta\omega d\mathcal{L} = \int_0^{\mathcal{L}}\frac{\delta\omega}{\omega_o}\omega_o d\mathcal{L}\ , \qquad 3.3.22$$

where ω_o is the rotational velocity of the equilibrium configuration.

Now, again making use of the conservation of angular momentum we see that

$$\omega r^2 \sin\theta = \text{const}\ . \qquad 3.3.23$$

Since we are only considering radial pulsation and thus $\delta\theta$ is zero, the variation (3.3.23) yields

$$\delta\omega r^2 \sin\theta + 2\omega r \delta r \sin\theta = 0\ . \qquad 3.3.24$$

This is equivalent to "conserving angular momentum in shells."

If we evaluate (3.3.24) at the equilibrium position, we obtain an expression of first order accuracy,

$$\frac{\delta\omega}{\omega_o} = -\frac{2\delta r}{r_o} = -2\xi_o e^{i\sigma t} \qquad 3.3.25$$

Substitution of this expression into equation (3.3.22) yields

$$2\delta \mathcal{T}_3 = -\int_o^{\mathcal{L}_o} \left[2 e^{i\sigma t} \xi_o \omega_o \right] d\mathcal{L}_o \qquad 3.3.26$$

If, for simplicity, we further assume that the rotational velocity is a constant throughout the configuration, we obtain a very simple form of the variation of the rotational energy.

$$2\delta \mathcal{T}_3 = -2 e^{i\sigma t} \xi_o \omega_o \mathcal{L}_o \qquad 3.3.27$$

where $\mathcal{L}_o$ is the total angular momentum for the system. Thus, only the variation of the magnetic energy remains to be determined.

In order to determine the variation of the total magnetic energy it is necessary to establish a coordinate system appropriate to the geometry of the field and to the geometry of the configuration. Although the configuration is spherically symmetric, the geometry of the magnetic field present is not known. Thus, we shall consider the variations in cartesian coordinates and later reduce our result to a form which is compatible with our previous results. Now the total magnetic energy of the configuration is defined (in c.g.s. units), as

$$\mathcal{M} = \int_o^M \frac{|\mathbf{H}|^2}{8\pi\rho} dm(r) \qquad 3.3.28$$

Thus, denoting the cartesian coordinates as x_1, x_2, and x_3, the variational form of the magnetic energy is

$$\delta \mathcal{M} = \frac{1}{4\pi} \int_o^M \frac{\mathbf{H} \cdot \delta\mathbf{H}}{\rho} dm(r) - \frac{1}{8\pi} \int_o^M |\mathbf{H}|^2 \frac{\delta\rho}{\rho^2} dm(r)$$

$$3.3.29$$

or, in cartesian coordinates

73

$$\delta \omega = \frac{1}{4\pi} \iiint \sum_i H_i \delta H_i \, dx_1 dx_2 dx_3$$

$$- \frac{1}{8\pi} \iiint |\mathbf{H}|^2 \frac{\delta\rho}{\rho} \, dx_1 dx_2 dx_3 \qquad 3.3.30$$

Although we have already obtained an expression for $\delta\rho/\rho$ in Section 3, due to the introduction of cartesian coordinates it is convenient to express this variation in terms of the variation of the coordinates n_i[3.5], namely

$$\frac{\delta\rho}{\rho} = - \sum_{i=1}^{3} \frac{\partial n_i}{\partial x_i} \, , \qquad 3.3.31$$

Before we can evaluate the expression for the variation of the magnetic energy, we must first determine the variation of the magnetic field δH_i. A rather lengthy argument[3.6] shows we can express this in terms of the coordinate variations, so

$$\delta H_i = \sum_j [H_j \frac{\partial n_i}{\partial x_j} - H_i \frac{\partial n_j}{\partial x_j}] \, . \qquad 3.3.32$$

If we substitute (3.3.32) and (3.3.31) into (3.3.29), we have

$$\delta \omega = \frac{1}{4\pi} \iiint \sum_i \sum_j H_i H_j \frac{\partial n_i}{\partial x_j} dx_1 dx_2 dx_3$$

$$- \frac{1}{4\pi} \iiint \sum_i \sum_j H_i^2 \frac{\partial n_j}{\partial x_j} dx_1 dx_2 dx_3$$

$$+ \frac{1}{8\pi} \iiint H^2 \sum_j \frac{\partial n_j}{\partial x_j} dx_1 dx_2 dx_3 \qquad 3.3.33$$

The second and third terms combine to yield

$$\delta \omega = \frac{1}{4\pi} \iiint \sum_i \sum_j H_i H_j \frac{\partial n_i}{\partial x_j} dx_1 dx_2 dx_3$$

$$+ \frac{1}{8\pi} \iiint |H|^2 \sum_j \frac{\partial n_j}{\partial x_j} dx_1 dx_2 dx_3 \qquad 3.3.34$$

Now, if we assume that η_i is only a function of x_i, then the sum on j in the first term collapses and the remaining terms result in

$$\delta \mathfrak{M} = \frac{1}{8\pi} \iiint \sum_i 2H_i^2 \frac{\partial \eta_i}{\partial x_i} - |H|^2 \frac{\partial \eta_i}{\partial x_i} \, dx_1 dx_2 dx_3 \qquad 3.3.35$$

At this point, it is appropriate to re-introduce the assumption concerning the nature of the variation η_i. It was earlier assumed that ξ_o was constant. The equivalent assumption for the η_i's is that

$$\eta_i = \text{const } x_i \qquad 3.3.36$$

Substitution of this explicit variation into (3.3.35) yields

$$\delta \mathfrak{M} = - \frac{\text{const.}}{8\pi} \iiint |H^2| \, dx_1 dx_2 dx_3 \qquad 3.3.37$$

Now, since we wish to consider the same type of pulsation from the η_i's, as we have assumed in the earlier section, we require that

$$\boldsymbol{n} = \delta \boldsymbol{r} \qquad 3.3.38$$

or

$$\frac{\boldsymbol{n}}{r} = \vec{\xi} = \xi_o e^{i\sigma t} \hat{\boldsymbol{r}} \qquad 3.3.39$$

Making use of our definition for η_i's, (3.3.36), we have

$$\text{const.} \left\{ \frac{x_i + x_j + x_k}{r} \right\} = \xi = \xi_o e^{i\sigma t} \qquad 3.3.40$$

This relation can only be true if the pulsation in the three coordinates (η_i) are in phase and of equal amplitude, and if

$$\text{const.} = \xi_o e^{i\sigma t} \qquad 3.3.41$$

Now, using the definition for the mass in a given volume in cartesian coordinates and the value for the constant in (3.3.37), we may re-write the variation of the magnetic energy as follows:

$$\delta \mathfrak{M} = - \xi_o e^{i\sigma t} \left\{ \frac{1}{8\pi} \int_o^M \frac{H^2}{\rho} dm(r) \right\} \qquad 3.3.42$$

Making use of 3.3.28 we may rewrite the variation in terms of the total magnetic energy of the equilibrium configuration

$$\delta \mathfrak{M} = - \xi_o e^{i\sigma t} \mathfrak{M}_o \quad . \qquad 3.3.43$$

Thus, we have obtained an expression for the last variation required to evaluate the variational form of the virial theorem (equations 3.3.7). We may, therefore, substitute equations (3.3.16), (3.3.27), and (3.3.43), into equation (3.3.7), and obtain

$$\frac{1}{2} \frac{d^2}{dt^2} (2e^{i\sigma t} \xi_o I_o) = - 2e^{i\sigma t} \xi_o \omega_o \mathcal{L}_o$$

$$+ 3(\gamma-1) (\xi_o e^{i\sigma t}) (\omega_o \mathcal{L}_o + \Omega_o + \mathfrak{M}_o)$$

$$- \xi_o e^{i\sigma t} \Omega_o - \xi_o e^{i\sigma t} \mathfrak{M}_o \quad . \qquad 3.3.44$$

Simplifying (3.3.44), we find that

$$\sigma^2 = \frac{-(3\gamma-4)(\Omega_o + \mathfrak{M}_o) + (5-3\gamma)\omega_o \mathcal{L}_o}{I_o} \quad . \qquad 3.3.45$$

Although we have made some strict assumptions in deriving equation (3.3.45), one should not feel that they are all of paramount importance. The assumption of constancy of ξ_o and γ were only made so that the resultant integrals could be integrated in terms of the original parameters. If necessary, these assumptions may be omitted and an integral expression similar to equation (3.2.18) may be derived. However, the work required to obtain this expression is non-trivial and in order for it to be useful one must have a detailed model in mind. One must also know the detailed geometry of the magnetic fields and of the star in order to evaluate the integrals that result. Also, to study the behavior of σ^2 a great deal of numerical work will be required. For purposes of studying the effects of changes in γ, Ω_o, M_o, ω_o, and $\mathcal{L}_o$, equation (3.3.43) will be quite adequate and is much easier to handle.

Equation (3.3.45) contains many aspects which one may check for 'reasonableness'. If we let η_o and ω_o be zero, then (3.3.45) becomes identical to the previously derived equation (3.2.19). Letting only ω_o be zero we obtain an expression identical to one arrived at by Chandrasekhar and Limber[5] (1954). If ω_o is non-zero, while $\mathcal{M}_o$ is zero, the expression is that of Ledeoux[1] (1945). It should not be surprising to find the magnetic energy entering in an additive manner to the gravitational energy. Both are potential energies and since the basic equations are scalar in nature, we should expect the final result to merely 'modify' the gravitational energy. However, the rotational energy is kinetic in nature and hence would not enter into the final result in the same manner as the magnetic energy.

Let us briefly investigate the effect upon σ^2, and hence on the pulsational period of the presence of magnetic and rotational energy. Since,

$$T = 2\pi/\sigma \qquad 3.3.46$$

an increase in σ indicates a decrease in the period and vice-versa. Now, since $\gamma > 4/3$ and the gravitational potential energy is defined as being negative, the first term in the numerator of (3.3.45) will be positive only if

$$|\Omega_o| > \mathcal{M} . \qquad 3.3.47$$

Thus, the introduction of magnetic fields only serves to reduce σ^2 and thereby lengthen the period of pulsation. However, the addition of rotational energy ($\omega_o \mathcal{L}_o$) will tend to increase σ^2 as long as $\gamma > 5/3$. When $\gamma = 5/3$, the introduction of rotation has no effect on the period of pulsation. If $\gamma < 5/3$, the influence of rotation is similar to that of magnetic energy.

4. Variational Form of the Surface Terms

In deriving the virial theorem, we noted earlier that the use of the divergence theorem yields some surface integrals which are generally ignored. Formally they may be ignored by taking the bounding surface of the configuration to be at infinity. However, in reality, this generally proves to be inconvenient for stars as they usually have a reasonably well-defined surface or boundary. For stars possessing general magnetic fields which extend beyond the surface, these surface contributions should be included. They are usually wished away by assuming they are small compared to the total magnetic energy arising from the volume integration. Although this may be true for simple fields in stars, it is unlikely to be true for other gaseous configurations such as flares and in any event a numerical estimate of their importance is far more re-assuring than an intuitive feeling. For this reason, let us consider the way in which these surface terms affect the variational formalism of the previous section.

To facilitate the calculations, we will assume the star is nearly spherical and the pulsations are also spherical. If the magnetic field is strong this will clearly not be the case and the full tensor virial theorem must be used. However, the simplicity generated by the use of the scalar virial theorem justifies the approach for purposes of illustration.

Let us begin by sketching the origin of the virial theorem as rigorously presented by Chandrasekhar[6]. The equations of motion for a gas with zero resistivity are

$$\rho \frac{d\mathbf{u}}{dt} = -\nabla p + \rho \nabla \Phi + \frac{1}{4\pi} (\nabla \times \mathbf{H}) \times \mathbf{H} \qquad 3.4.1$$

Employing the identity $(\nabla \times \mathbf{H}) \times \mathbf{H} = (\mathbf{H} \cdot \nabla)\mathbf{H} - 1/2\ \nabla(\mathbf{H} \cdot \mathbf{H})$ and taking the scalar product of 3.4.1 with the position vector $\mathbf{r}$ then integrating over all space enclosed by the bounding surface, we get

$$\int_V \rho\, \mathbf{r} \cdot \frac{d\mathbf{u}}{dt}\, dV = -\int_V \mathbf{r} \cdot \nabla P\, dV + \int_V \rho\, \mathbf{r} \cdot \nabla \Phi\, dV$$
$$+ \frac{1}{4\pi} \int_V \mathbf{r} \cdot (\mathbf{H} \cdot \nabla)\mathbf{H}\, dV - \frac{1}{8\pi} \int_V \mathbf{r} \cdot \nabla(H^2)\, dV \ . \qquad 3.4.2$$

As in Section 3, this becomes

$$\frac{1}{2}\frac{d^2 I}{dt^2} - 2T = 3(\Gamma-1)\mathcal{U} + \Omega + \mathcal{M} - \int_S P_o \mathbf{r} \cdot d\mathbf{s}$$

$$-\frac{1}{8\pi}\int_S H_o^2 \mathbf{r}_o \cdot d\mathbf{s}$$

$$+\frac{1}{4\pi}\int_S (\mathbf{r}_o \cdot \mathbf{H}_o)(\mathbf{H}_o \cdot d\mathbf{s}) \qquad 3.4.3$$

where P_o and $\mathbf{H}_o$ are the gas pressure and magnetic field present at the surface $\mathbf{r}_o$. It is the behavior of the three integrals in 3.4.3 that will interest us as hopefully the remaining terms are by now familiar.

Consider first the effect of a pulsation on the surface term arising from the pressure by taking the variations of the surface pressure integral.

$$\delta\int_S P_o \mathbf{r}_o \cdot d\mathbf{s} = \int_S \delta P \; \mathbf{r}_o \cdot d\mathbf{s} + \int_S P\delta \mathbf{r}_o \cdot d\mathbf{s}$$

$$+ \int_S P_o \mathbf{r}_o \cdot d(\delta \mathbf{s}) \qquad 3.4.4$$

For radial variations only

$$d(\delta s) = 2r_o \delta r_o \sin\theta \, d\theta d\phi = 2\xi r_o^2 \sin\theta \, d\theta d\phi = 2\xi dS ;$$

$$3.4.5$$

where as in Section 2, $\xi = \delta r/r$. In Section 2 (equation N 3.2.13 and 3.2.14), we already have shown that for adiabatic pulsations

$$\frac{\delta P}{P} = \gamma \frac{\delta \rho}{\rho} = -\gamma(3\xi + r d\xi/dr) . \qquad 3.4.6$$

Combining 3.4.5 and 3.4.6 with 3.4.4, we get

79

$$\delta \int_S P_o \mathbf{r}_o \cdot d\mathbf{s} = 3(1-\gamma) \int_S \xi P_o \mathbf{r}_o \cdot d\mathbf{s} \qquad 3.4.7$$

$$- \gamma \int_S \left(r_o \frac{d\xi}{dr} \bigg|_{r_o} \right) \mathbf{r}_o \cdot d\mathbf{s}$$

Earlier we assumed that ξ was constant throughout the star and hence its derivative vanished. Here, we only require the derivative to vanish at the surface in order to simplify 3.4.7 to get

$$\delta \int_S P_o \mathbf{r}_o \cdot d\mathbf{s} = 3(1-\gamma)\xi \int_S P_o r_o dS. \qquad 3.4.8$$

Now consider the variation of the two magnetic integrals in (3.4.3).

$$\delta \left\{ \frac{1}{8\pi} \int_S 2(\mathbf{r}_o \cdot \mathbf{H}_o)(\mathbf{H}_o \cdot d\mathbf{s}) - \frac{1}{8\pi} \int_S H_o^2 \mathbf{r}_o \cdot d\mathbf{s} \right\}$$

$$= \frac{1}{8\pi} \int_S 2(\delta \mathbf{r}_o \cdot \mathbf{H}_o)(\mathbf{H}_o \cdot d\mathbf{s}) + \frac{1}{8\pi} \int_S 2(\mathbf{r}_o \cdot \delta \mathbf{H}_o)(\mathbf{H}_o \cdot d\mathbf{s})$$

$$+ \frac{1}{8\pi} \int_S 2(\mathbf{r}_o \cdot \mathbf{H}_o)(\delta \mathbf{H}_o \cdot d\mathbf{s}) + \frac{1}{8\pi} \int_S 2(\mathbf{r}_o \cdot \mathbf{H}_o)(\mathbf{H}_o \cdot d\delta \mathbf{s})$$

$$+ \frac{1}{8\pi} \int_S 2(\mathbf{H}_o \cdot \delta \mathbf{H}_o)(\mathbf{r}_o \cdot d\mathbf{s}) - \frac{1}{8\pi} \int_S H_o^2 \delta \mathbf{r}_o \cdot d\mathbf{s}$$

$$- \frac{1}{8\pi} \int_S H_o^2 \mathbf{r}_o \cdot d(\delta \mathbf{s}).$$

$$3.4.9$$

This truly horrendous expression does indeed simpify [3.7] by using

$$\delta \mathbf{H} = -2\xi \mathbf{H}_o \qquad 3.4.10$$

Using this result, equation 3.4.5 and the definition of ξ, equation 3.4.9 becomes:

$$\delta Q_m = -\frac{\xi}{8\pi} \left\{ \int_S 2(\mathbf{r}_o \cdot \mathbf{H}_o)(\mathbf{H}_o \cdot d\mathbf{s}) - \int_S H_o^2 \mathbf{r}_o \cdot d\mathbf{s} \right\} \quad 3.4.11$$

where Q_m stands for the original magnetic surface term that appears in equation 3.4.3. Thus, the variation of the surface terms can be represented as

$$\left. \begin{array}{r} \delta Q_m = -\xi Q_m \\ \\ \delta Q_p = 3(1-\gamma)\xi Q_p \end{array} \right\} \quad 3.4.12$$

and

If we assume a linear or homologous pulsation, as was done in the previous two sections, then the expression for the pulsational frequency (equation 3.3.45), becomes

$$\sigma^2 = -\frac{(3\gamma-4)(\Omega_o + \mathcal{M}_o) + (5-3\gamma)\omega_o \mathcal{L}_o + (3\gamma-1)Q_p - Q_m}{I_o} \quad 3.4.13$$

since $\gamma > 4/3$ the contribution of the surface pressure term is such as to increase σ^2 and thereby improve the stability of the system. Basically, this results because an unstable system will have to do work against the surface pressures either in expanding or contracting the surface. This energy is thus not available to feed the instability. The situation is not as obvious for the magnetic contribution Q_m, since Q_m is the difference between two positive quantities. Thus, the result depends entirely on the geometry of the field.

The effect of the field geometry can be made somewhat clearer by considering a spherical star so that the radius vector is parallel to the surface normal. Under these conditions Q_m becomes

$$Q_m = \frac{1}{8\pi} \int_S [2(\hat{\mathbf{H}}_o \cdot \hat{\mathbf{r}})^2 - 1] H_o^2 \, rds$$

$$= \frac{1}{8\pi} \int_S \cos 2\beta \, H_o^2 \, rds \quad 3.4.14$$

where β is the local angle between the field and the radius

81

vector. Thus, the average of cos 2β weighted by H_o^2 over the surface will determine the sign of Q_m. In any event, it is clear that

$$|Q_m| < \frac{1}{8\pi} \int_S H_o^2 \, r_o \, ds = \frac{1}{2} R_o^3 \overline{H_o^2} \ . \qquad 3.4.15$$

It is worth noting that in the case where the magnetic field increases slowly with depth, this term can be of the same order of magnitude as the internal magnetic field energy and must be included. Furthermore, whether or not the local contribution to Q_m is positive or negative depends on whether or not the local value of β is greater or less than π/4. Since a positive value of Q_m increases the value of σ^2, fields exhibiting a local angle to the radius vector greater than π/4 tend to stabilize the object, whereas more radial fields enhance the instability. This simply results from the fact that a radial motion will tend to compress fields more nearly tangential to the motion than 45°, thereby removing energy from the motion. Conversely, more nearly radial fields will tend to feed the perturbation leading to a decrease in stability.

One thing becomes immediately clear from this discussion. If Q_m is an important term in 3.4.13, radial pulsation will not occur. Since a magnetic field cannot exhibit spherical symmetry, the departures from symmetry will yield a variable "restoring force" over the surface inferring that non-radial modes will be excited. In this case the tensor virial theorem must be used and the field geometry known.

Lastly, for purposes of simplicity, we have assumed no coupling between the gas pressures and magnetic pressures. Unless the system is rather bizarre, the gas will be locally relaxed on a time scale less than the pulsation period and hence the two cannot be treated independently. This assumption was made merely for the sake of simplicity and doesn't affect the illustrative aspects of the effects. However, unless Q_m and Q_p are comparable the coupling between the two will be weak and we may expect 3.4.13 to give good quantitative results.

5. The Virial Theorem and Stability

In the last section, I alluded to the effects that the surface terms have on the stability of the system being considered. This concept deserves some amplification as it represents one of the most productive applications of the virial theorem. However, before embarking on a detailed development of the virial theorem for this purpose, it is appropriate to review the use of the word stability itself.

When inquiring into the meaning of the word, it is customary to consult a dictionary. This approach provides the following definition:

> Stability: "That property of a body which causes it, when disturbed from a condition of equilibrium or steady motion, to develop forces or moments which tend to restore the body to its original condition."

This definition is subject to several interpretations and serves to illustrate the danger of consulting an English dictionary to learn the meaning of a technical term.

The word stability is usually associated with the word equilibrium. This is primarily because the concept of stability normally is first encountered during the study of statistics. However, there are many dynamical situations, which are not equilibrium situations, that even the most skeptical person would call stable. One of the most obvious examples to an astronomer, are the stars themselves. Not all stars would be regarded as stable, but certainly most of the main sequence stars are. Since stars are not really equilibrium configurations, but rather steady state configurations, we see that we must extend our conceptualization of stability to include some dynamical systems.

The normal definition of equilibrium requires that the sum of all forces acting on the system is zero. This concept may be broadened to dynamical systems if one requires that the generalized forces (Q_i) acting on the systems are zero. Here the concept of the generalized force may be most simply stated as

$$Q_i = \sum_j \mathbf{F}_j \cdot \frac{\partial \mathbf{r}_j}{\partial q_i} \quad . \qquad 3.5.1$$

where $\mathbf{F}_j$ represents the physical forces of the system acting on the jth particle and the q_i's represent any set of linearly independent 'coordinates' adequate to describe the system. In a conservative system all the forces are derivable from a potential Φ. Thus, the generalized forces may be written as

$$Q_i = - \sum_j \nabla_j \Phi \cdot \frac{\partial \mathbf{r}_j}{\partial q_i} = - \sum_j (\frac{\partial \Phi}{\partial \mathbf{r}_j}) \hat{\mathbf{r}}_j \cdot (\frac{\partial \mathbf{r}_j}{\partial q_i})$$

$$= - \frac{\partial \Phi}{\partial q_i} .$$

3.5.2

Thus, saying the generalized forces must vanish is equivalent to saying the potential energy must be in extremum.

$$Q_i = - (\frac{\partial \Phi}{\partial q_i}) \bigg|_{q_i = q_i(o)} = 0$$

3.5.3

Now, in terms of this definition of equilibrium we may proceed to a definition of stable equilibrium. If the potential extremum implied by (3.5.3) is a minimum then the equilibrium is said to be stable. The conditions thus imposed on the potential are simply

$$\left(\frac{\partial^2 \Phi}{\partial q_i^2}\right) \bigg|_{q_i = q_i(o)} > 0 .$$

3.5.4

In order to see that this definition of stability is consistent with our dictionary definition, consider the following argument. Suppose a system is disturbed from equilibrium by an increase in the total energy dE above the total energy at equilibrium. If Φ is a minimum, any disturbances from equilibrium will produce an increase in the potential energy. Since the conservation of energy will apply to the system after the incremental energy dE has been applied, the kinetic energy must decrease. This implies that the velocities will decrease for all particles and eventually become zero. Thus, the motion of the system will be bounded (Note: the bound may be arbitrarily large). If, however, the departure from equilibrium brings about a decrease in the potential energy, then the velocities may increase without bound. We would certainly call such motion unstable motion.

However, simple and clear-cut as this definition of stability may seem, it is still inadequate to serve the needs of mathematical physicists in describing the behavior of systems of particles. Thus it is not uncommon to find modifying adjectives or compound forms of the word "stable" appearing in the literature. A few common examples are: Secular stability, global stability, quasi-stable, bi-stable, and over-stable. These terms are usually used without definition in the hope that the reader will be able to discern the correct meaning from the context. The introduction of these modifiers as often as not arises from the mode of analysis used to describe the system. It is a common practice to examine the response of the system to a continuous spectrum of perturbations. If any of these perturbations grow without bound the system is said to be unstable. This would seem in full accord with our dictionary definition and thereby wholly satisfying. Unfortunately, one is rarely able to calculate the response in general. It is usually necessary to linearize the equations describing the system in order to solve them. Analysis of this type is called linear stability theory, and is actually the basis for most stability criterion. Thus, when analyzing a system not only must one correctly carry out the stability analysis, he must also decide on the applicability of the analysis to the system.

Recently, it has been quite fashionable to use the virial theorem as the vehicle to carry out linearized normal mode analysis of systems in order to determine their state of stability. However, the determination of a system state of stability seems to have inspired Jacobi to develop the n-body representation of Lagrange's identity from which it is a short step to the virial theorem.

To see now how closely tied the virial theorem is connected to stability, let us summarize some of Jacobi's arguments.

In Chapter 1, (Equation 1.4.12), we arrived at simple statements of Lagrange's identity for self-gravitating systems as:

$$\frac{1}{2} \frac{d^2 I}{dt^2} = 2T + \Omega \qquad 3.5.5$$

One could say with some confidence that if $d^2I/dt^2 > 0$ for all t the system would have to have at least one particle whose position coordinates increased without bound. That is to say, the system would be unstable. However, since both T and Ω vary with time it would be difficult to say something a priori about d^2I/dt^2 from Lagrange's identity alone. Thus, Jacobi employed the constancy of the total energy (i.e., $E = T + \Omega = $ const.), and the fact that for self-gravitating systems $\Omega > 0$ to modify 3.5.5 to give:

$$\frac{1}{2} \frac{d^2I}{dt^2} = 2E - \Omega > 2E \qquad 3.5.6$$

So, if $E > 0$, $d^2I/dt^2 > 0$ and the system is unstable. This is known as Jacobi's stability criterion and provides a sufficient (but not necessary) condition for a system to be called unstable.

It is the constancy of E with time that make this a valuable criterion for stability and is the reason Jacobi used it. There is another approach to the problem of temporal variation which was not available to Jacobi. In Chapter 2 we discussed the extent to which time averages of quantities may be identified within their phase averages, through use of the Ergodic theorem. Integrating 3.5.5 over some time t_o, we get

$$\frac{1}{t_o} \left[\frac{dI_o}{dt} \right] \bigg|_o^{t_o} = 2\overline{T} + \overline{\Omega} \qquad 3.5.7$$

If the system is to remain bounded, dI/dt must always be finite. If the system is to be always stable, then the limit of the left-hand side must tend to zero. Furthermore, the time averages on the right-hand side of 3.5.7 will tend to phase averages if the system is Ergodic. Thus

$$2 <T> + <\Omega> = 0 \qquad 3.5.8$$

constitutes a stability criterion for Ergodic systems. That is, 3.5.8 must be satisfied in a stable Ergodic system and failure of 3.5.8 is sufficient for instability. It is not uncommon to find the statement in the literature that $2T + \Omega > 0$ insures the instability of a system citing the virial theorem as the justification. Actually, it is Lagrange's identity that is the relevant expression and it only guarantees that at the moment the system is acceleratively expanding. What is really meant is that if $T + \Omega > 0$ the system is indeed unstable as this is just a statement of Jacobi's stability criterion concerning the total energy of the system.

Now, let us turn the applications of the variational form of the virial theorem to stability, keeping in mind that the variational approach is essentially a first order or linearized analysis.

The majority of Chapter 3 has been devoted to obtaining expressions for the frequency of a pulsating system. We obtained a value for the square of the frequency in terms of the equilibrium energies of the configuration. However, these expressions could be neither positive or negative. In the earlier sections, we discussed only the meaning of the positive values, as negative squares of frequencies had no apparent physical meaning. Let us look again at the nature of the assumed pulsation in order to further investigate the meaning of these pulsation frequencies. In Section 2 (equations 3.2.12), we assumed that the pulsation would be periodic and of the form

$$\frac{\delta r}{r} = \xi_o e^{i\sigma t} \quad . \qquad 3.5.9$$

where σ was the frequency of pulsation and ξ_o did not depend on time. Now, if we make the formal identification between σ and the period and take σ to be purely imaginary, we may write

$$\sigma = \pm 2\pi i/t_o \quad , \qquad 3.5.10$$

where t_o is a real number. Combining (3.5.10) and (3.5.9), we have

$$\frac{\delta r}{r} = \xi_o e^{\pm 2\pi t/t_o} \quad . \qquad 3.5.11$$

Thus, the pulsation becomes exponential in nature. If the sign of σ is negative, then the sign of the exponential in 3.5.11 will be positive and the pulsation will grow without bound with a rate of growth determined

87

by t_o.

One might be tempted to choose the negative sign of 3.5.11 saying that the system is stable as the pulsation will die out, even though $\sigma^2 < 0$. This would be wrong. All classical equations of dynamical symmetry exhibit full-time symmetry, thus those solutions which damp out in the future were unstable in the past and vice-versa. A specific solution would be fully determined by the boundary conditions at $t = 0$. Further, we assumed that a full continuum of perturbations are present, resulting from small but inevitable, departures from perfection produced by statistical fluctuations. Thus, if there exists even one mode with $\sigma^2 < 0$, the instability associated with that mode will grow without bound. Therefore, this becomes a sufficient condition for the system to be unstable in the strictest sense, of the word

$$\sigma^2 < 0 . \qquad 3.5.12$$

It is also worth noting that this criterion applies to the entire system, and thus is a "global" stability condition. However, it can be made into a local condition by taking an infinitesimal volume and including the surface terms discussed in Section 4.

Now, let us see what implications this analysis has for the stability of stars. In Section 2, we established an expression for the pulsation frequency of a gravitating gas sphere (equation 3.2.19). Now, applying the instability criterion (3.5.12), we see that the sphere will become unstable, when

$$-\frac{(3\gamma-4)\Omega_o}{I_o} < 0 . \qquad 3.5.13$$

Since the moment of inertia (I_o) is intrinsically positive, while the gravitational potential energy is intrinsically negative, equation (3.5.13) becomes

$$\left. \begin{array}{c} (3\gamma-4) < 0 \\ \\ \gamma < 4/3 \end{array} \right\} \qquad 3.5.14$$

or

88

Thus, a star will become unstable when γ is less than 4/3. This is the familiar instability criterion demonstrated by Chandrasekhar in <u>Stellar Structure</u>.[7] He further demonstrates that a gas with γ equal to 4/3 corresponds to a gas where the total pressure is entirely due to radiation. If we consider the other stability criterion (equation 3.5.14), we can see that a necessary condition for stability of a homogeneous non-rotating gas sphere is

$$\gamma > 4/3 . \qquad 3.5.15$$

Thus, having extracted as much information as possible from the pulsation expression developed in Section 2, let us turn to the more general formulae resulting from our analysis in Section 3. Remember the final expression for the pulsational frequency was

$$\sigma^2 = \frac{-(3\gamma-4)(\Omega_o + \mathcal{M}_o) + (\delta-3\gamma)\omega_o \mathcal{L}_o}{I_o} . \qquad 3.5.16$$

Consider first a gas sphere which is not rotating, but which has a magnetic field. Equation (3.5.16) becomes then

$$\sigma^2 = \frac{-(3\gamma-4)(\Omega_o + \mathcal{M}_o)}{I_o} . \qquad 3.5.17$$

Substituting this into the instability criterion (3.5.12), we have

$$(\Omega_o + \mathcal{M}_o)(3\gamma-4) < 0 . \qquad 3.5.18$$

Now, if we assume $\gamma > 4/3$, we have a sufficient condition for instability due to the presence of magnetic energy as follows:

$$\mathcal{M}_o > |\Omega_o| . \qquad 3.5.19$$

We may obtain a crude estimate of the magnitude of the magnetic fields necessary to disrupt a star in the following manner. The gravitational potential energy for a sphere of uniform density is just

$$\Omega = -\frac{3}{5} \frac{GM^2}{R} , \qquad 3.5.20$$

while the magnetic energy is

$$\mathfrak{M} = \frac{1}{8} \iiint |\mathbf{H}|^2 \, dx_1 dx_2 dx_3 = \frac{R^3 \overline{|\mathbf{H}|^2}}{6} . \qquad 3.5.21$$

Combining (3.5.20) and (3.5.21), we see that the root mean square value of the magnetic field sufficient to disrupt a uniformly dense sphere is

$$\sqrt{\overline{|H|^2}} > 2 \times 10^8 \, M/R^2 \text{ gauss} \qquad 3.5.22$$

where M and R are given in solar units. Thus, for a main sequence A star with $M = 4M_\odot$ and $R = 5R_\odot$, we have

$$H_{rms} > 3 \times 10^7 \text{ gauss} \qquad 3.5.23$$

However, for a star like VV Cephei with $M = 100 \, M_\odot$, and $R = 2600 \, R_\odot$, we have

$$H_{rms} > 3000 \text{ gauss} . \qquad 3.5.24$$

We may conclude from these arguments that for a main sequence star, an extremely large magnetic field would be sufficient to cause the star to become unstable. However, for an unusually large star, the required field becomes much smaller. In the case of VV Cephei, Babcock has measured a field ranging from +2000 to -1200 gauss. Thus, it would appear that VV Cephei is on the verge of being magnetically unstable. One might argue that our crude estimates of Ω are so crude as to be meaningless due to the large central concentration of the mass in giant stars. However, it should be pointed out that the magnetic field one can observe is, of necessity, a surface field and, therefore, provides us with a lower limit on the magnetic energy. Thus, we may have some hope that our limiting field values are not too far from being realistic.

It is interesting to note that the instability criterion (3.5.18) permits the existence of a gas with $\gamma > 4/3$ providing the magnetic energy exceeds the gravitational energy. Indeed, the stability criterion (3.5.18) would require a necessary condition for the stability of any configuration where $\mathfrak{M} > |\Omega|$ that γ be less than 4/3. However, it is also true that the physical meaning of a gas having a $\gamma < 4/3$ is a little obscure.

If we now consider a rotating configuration with no magnetic field, equation (3.5.14) combined with the stability criterion (3.5.12) becomes

$$(5-3\gamma)\omega_o \mathcal{L}_o > \Omega_o (3\gamma-4) . \qquad 3.5.25$$

If we restrict γ to be less than 5/3 we have

$$\omega_o \mathcal{L}_o > \frac{(3\gamma-4)\Omega_o}{(5-3\gamma)} . \qquad 3.5.26$$

Since Ω_o is intrinsically negative, we see that the stability condition will always be satisfied with any ω_o. Thus, for all known stars the stability criterion for rotation is not particularly useful.

However, all this is not meant to imply that the rotational terms are unimportant. Indeed, Ledoux[1] has shown that rotational velocities encountered in stars may lead to a variation in the pulsational period by as much as 20%.

Let us briefly consider the instability criterion when both magnetic and rotational energy are present. This may be obtained by combining 3.5.15 and 3.5.12, yielding

$$(3\gamma-4)(\Omega_o + \mathfrak{M}_o) - (5-3\gamma)\omega_o \mathcal{L}_o > 0 . \qquad 3.5.27$$

As before, this condition may never be satisfied unless $|\Omega_o| > \mathfrak{M}_o$. However, even in the event that $|\Omega_o| > \mathfrak{M}_o$, the condition may still not be satisfied because of the presence of the rotational term. Thus, it is evident that if $4/3 < \gamma < 5/3$, then the presence of rotation will help stabilize stars. This result is certainly not intuitive. A physical explanation of the result might be supplied by the following argument.

Consider a pulsating configuration containing both rotational and magnetic energy. As the system expands or contracts, a certain amount of energy will be required to slow down or speed up the rotation in order to keep the angular momentum constant. This energy must be supplied by the kinetic energy of the gas itself, and, since this is supplied by the potential energies present, ultimately must come from the gravitational and magnetic energies. Therefore, the amount of this energy transferred from the magnetic and gravitational energies will depend on γ. Also since the gravitational and magnetic energies must supply this energy to the rotation, the energy is no longer available to "feed" the pulsation and disrupt the star.

We may now ask what sort of increase in the maximum magnetic field can this additional rotational "stability" supply. From our previous investigation with the rotational stability criteria we might expect the result to be small. That is, since the rotational stability criteria did not supply us with as important a result as did the magnetic instability, we would expect the effects of rotation to be small compared with the magnetic energy. If one considers a uniform model with $\gamma = 3/2$, rotating at critical velocity, he will find the magnetic field may only be increased by about 0.3% before instability will again set in. Thus, even though stability is increased by the presence of rotation, it is not increased a great deal.

It is appropriate at this point to make some comments regarding all of the stability criteria relating to the stability of radial pulsations. It would have been more correct to employ the integral form of the expressions for the frequency of pulsation. However, the result one would obtain by using the integral expression and a specific model would only differ in degree from those derived here. It is hoped that the degree of differences would not be large.

There is one respect in which the differences between the derived criteria and the 'correct' ones may result in a difference in kind. It must be remembered that the expressions developed for the pulsational frequencies were based on a first order theory. Hence, so are the stability criteria developed in this section. However, the conditions at which one wishes to apply an instability criteria are generally such that the second, and higher, order terms are not small and should not be neglected. Chandrasekhar and Fermi[3] have shown that a sphere under the influence of a strong dipole field will tend to be "flattened" in much the same way as it will be by rotation. Once the spherical symmetry has been destroyed, either by the presence of a strong, magnetic field or rapid rotation, the concept of radial pulsation becomes inconsistent. As mentioned before, analysis of such systems would require the use of the tensor virial theorem and considerable insight into the types of perturbations to employ.

I would be remiss if I left the subject of the virial theorem and stability without some discussion of the recent questions raised with regard to the appropriateness of the approach. To me, these questions appear to be partly substantive and partly semantic and revolve largely around one of these modifiers mentioned earlier, namely, secular stability. So far, our discussion has been restricted to problems involving dynamical stability about which there seems to be little argument. A dynamically unstable system will disintegrate exponentially, usually on a time scale related to the hydrodynamical time scale for the system. Such destruction is usually so unambiguous that no complications arise in the use of the word unstable. Such is not the case for the term secular stability.

The notion of secular stability involves the response of the system to small dissipative forces, such as viscosity and thus must depend to some extent, on the nature of those forces. Time scales for development of instabilities will be governed by the forces and hence may be very long. Perhaps one of the clearest contemporary discussions of the term is given by Hunter[8] who notes that there is less than universal agreement on this meaning of the term. He points out that difficulties arise in rotating systems resulting from the presence of the coriolis forces, which lead to a clear distinction between dynamically and secularly stable systems. As we saw in Chapter 2, Section 5, the terms associated with the coreolis forces can be made to vanish by the proper choice of a coordinate frame and they would appear to play no role in the energy balance of the system. However, their variation does not vanish and hence they will affect the pulsational analysis. Since globally the forces are conservative the first result is not surprising and since radially moving mass in a rotating frame must respond to the conservation of angular momentum, neither is the second.

Now if dissipative forces are present such as viscosity then it may be possible to redistribute local angular momentum while conserving it globally so that no equilibrium configuration is ever reached. In addition, except for the global constraint on the total angular momentum, no constraints are placed on the transfer of energy from the rotational field to the thermal field. Indeed the presence of dissipative forces guarantees that this must happen. Thus, instabilities associated with these forces might exist which would otherwise go undetected. This line of reasoning demonstrates a qualitative difference between the cases of uniform rotation and differential rotation in that in the former dissipative forces will be inactive and the analysis will be appropriate while in latter cases they must be explicitly included. This point is central to a lengthy series of papers[9,10,11,12,13] by Ostriker and others, which discuss the stability of a variety of differentially rotating systems. However, the majority of these papers clearly state that the authors are dealing with systems with zero viscosity and so the problem is not one of the accuracy of the analysis but rather of the applicability of the analysis to physical systems. In practice, the viscosity of the gas in most stars is so extremely low that the time scales for the development of instability arising from viscosity driven instabilities will be very long.

One cannot hope to untangle in a few paragraphs a controversy which has taken more than a decade to develop and at a formal mathematical level is quite subtle. However, it is worth noting that recent[14] statements which essentially say that the tensor virial approach to stability is wrong do nothing to clarify the situation. The presence of dissipative forces can be included in the equations of motion and thus in the resulting tensor representation of Lagrange's identity. The resulting stability analysis would then correctly reflect the presence of these forces and thus be dependent on their specific nature. Insofar as the time averaged form of Lagrange's identity, which is technically the virial theorem, is used, the arguments of Milne[15] as presented in Chapter 1 still apply. The presence of velocity dependent forces does not affect the virial theorem unless those forces stop or destroy the system during the time over which the average is taken. At this level assailing the virial theorem is as useful an enterprise as denying the validity of a conservation law.

6. Summary

In this chapter we have explored the results of applying a specific analytical technique to the virial theorem. As in other chapters, we began with the simple and moved to the more complex. Having discussed the implications of the variational approach to the virial theorem we moved to develop the explicit form for the simple scalar theorem appropriate for self-gravitating systems. We re-created the pulsational formula (equation 3.2.19) originally due to Ledoux. One implication of this result is that the fundamental mode of oscillation depends only on the square root of the density and when coupled with the stability criterior in Section 5, leads immediately to the Jean's stability criterion. This is not surprising as both results have as the derivational origin the same concept (i.e., the equations of motion). However, it is assuring when a different approach yields results already well accepted.

In Section 3 we expanded the variational approach to include the effects of magnetic fields and rotation. In spite of many distractions dealing with the variation of magnetic fields, etc., the influence of these added features on the pulsation frequency and hence stability became clear. Rotation can either enhance or reduce the stability of a configuration depending on whether or not the value of γ for the gas permits net energy to be fed to the pulsation. The influence of an internal magnetic field is to de-stabilize the star for all realistic values of γ. However, the effect of a surface field proves to be more complex. Here the result depends critically on the geometry of the field.

In the last section we dealt briefly with the overall question of stability and showed explicitly how the virial theorem provided an excellent basis for a linear stability analysis of a symmetric system. Throughout the chapter we confined ourselves to spherically symmetric systems exhibiting radial pulsations only. As mentioned, this is inappropriate when considering either rotation or magnetic fields as neither can exhibit spherical symmetry and thus one would expect non-radial oscillation to be excited. However, unless the field energies become quite large one would expect the pulsational frequencies not to differ greatly from the purely radial theory.

This line of reasoning becomes particularly dangerous when one turns to a discussion of stability. First the interesting situations of marginal stability are liable to involve substantial magnetic fields or rapid rotation. If these aspherical properties are large, the departure from spherical systems of the mass destribution will also be large invalidating every aspect of the analysis. In addition, for the stability analysis to be valid all possible modes of perturbation must be included. Limiting oneself to only the radial modes is to invite a misleading result.

Fortunately the techniques for dealing with these problems exist and have been developed here as well as the literature. The tensor virial theorem as is presented in Chapter II, Section 1 allows one to follow perturbations in independent spatial coordinates. In principle, a complete variational analysis of perturbations to all independent spatial coordinates will allow one to compute the non-radial as well as radial modes of oscillation and thereby obtain a much more secure analysis of the system's stability.

Notes to Chapter 3

3.1 Remember the defining expression for the gravitational potential energy is

$$\Omega = - G \int_o^M \frac{m(r)dm(r)}{r} \ .$$ N3.1.1

We may again use the fact that the variation of $m(r)$ and $dm(r)$ are both zero, to obtain

$$\delta\Omega = G \int_o^M \frac{\delta r}{r^2} m(r)dm(r)$$ N3.1.2

which can be written as an energy integral by noting from (N3.1.1) that

$$\frac{d\Omega}{dm(r)} = - \frac{Gm(r)}{r} \ .$$ N3.1.3

Evaluating the above expression at $r = r_o$ and using the result in (3.2.11) to first order accuracy we get

$$\delta\Omega = - \int_o^M \frac{\delta r}{r_o} d\Omega.$$ N3.1.4

3.2 We may write the total energy as the sum of two energies $\mathcal{T}_1$ and $\mathcal{T}_2$, where $\mathcal{T}_1$ is the kinetic energy due to the mass motion of the gas arising from the pulsations themselves. Now the total kinetic energy of mass motion is given by

$$\mathcal{T}_1 = \frac{1}{2} \int_o^R 4\pi r^2 \rho \ (\frac{dr}{dt})^2 dr = \frac{1}{2} \int_o^M (\frac{dr}{dt})^2 dm(r) \ .$$ N3.2.1

Thus, the variation of $\mathcal{T}_1$ is

$$\delta\mathcal{T}_1 = \int_o^M (\frac{dr}{dt}) \frac{d(\delta r)}{dt} dm(r) \ .$$ N3.2.2

However, from the definition of δr we see that

$$r = r_o + \delta r \ .$$ N3.2.3

Since the equilibrium point r_o cannot vary with time by definition, equation (N3.2.2) becomes

$$\delta\mathcal{T}_1 = \int_o^M [\frac{d}{dt} (\delta r)]^2 dm(r) \ .$$ N3.2.4

The largest term in the integral of (N3.2.4) is second order in δr and may be neglected with respect to the first order terms of (3.2.13), and (3.2.6). Thus, to first order we have

$$\delta T = \delta\mathcal{T}_1 + \delta\mathcal{T}_2 \simeq \delta\mathcal{T}_2 \ .$$ N3.2.5

Now consider the kinetic energy of a small volume of an ideal gas.

$$d\mathcal{T}_2 = \frac{3}{2} Nk\mathcal{T} \ dV \ .$$ N3.2.6

However, the gas pressure is given by $P_g = Nk\mathcal{T}$ and $dm(r) = \rho dV$. Therefore,

$$d\mathcal{T}_2 = \frac{3}{2} \frac{P_g}{\rho} dm(r) \ .$$ N3.2.7

Thus, twice the total kinetic energy of the gas sphere arising from thermal sources is

$$2\mathfrak{I}_2 = 3 \int_o^M \frac{P_g}{\rho} \, dm(r) \ . \qquad \text{N3.2.8}$$

Now neglecting radiation pressure, so that the total pressure is equal to the gas pressure, and remembering that the variation of dm(r) is zero, we have

$$2\delta T \simeq 2\delta\mathfrak{I}_2 = 3 \int_o^M \delta(\frac{P}{\rho}) \, dm(r) \qquad \text{N3.2.9}$$

We shall now assume that the pulsations are adiabatic so that

$$\frac{\delta P}{P} = \gamma \, \frac{\delta\rho}{\rho} \qquad \text{N3.2.10}$$

where γ is the ratio of specific heats c_p/c_v. Now

$$\delta(P/\rho) = \frac{\rho\delta P - P\delta\rho}{\rho^2} = \frac{(\frac{\delta P}{P})\rho - \delta\rho}{\rho^2/P} = \frac{P}{\rho^2}[(\gamma-1)\delta\rho] \qquad \text{N3.2.11}$$

Therefore, again evaluating at the equilibrium position, substuting into equation (N3.2.9), and keeping only terms up to first order, we have

$$2\delta T \simeq 3 \int_o^M \frac{P_o}{\rho_o} (\gamma-1) \frac{\delta\rho}{\rho_o} \, dm(r) \ . \qquad \text{N3.2.12}$$

3.3   In our attempt to obtain an expression for $\delta\rho/\rho$ and thereby determining the variation of the kinetic energy, we shall invoke the following argument.

From the conservation of mass, we have

$$\delta m(r') = 0 = \delta \int_o^{r'} 4\pi r^2 \rho \, dr = \int_o^{r'} 4\pi(2r\delta r)\rho \, dr \qquad \text{N3.3.1}$$
$$+ \int_o^{r'} 4\pi r^2 \delta\rho \, dr + \int_o^{r'} 4\pi r^2 \rho \, d(\delta r) \ .$$

Re-writing (3.2.28) and evaluating at $r = r_o$, we have

$$\int_o^{r'} r_o^2 \delta\rho \, dr_o = - \int_o^{r'} \rho_o r_o^2 [\frac{2\delta r \, dr_o}{r_o} + d(\delta r)] \ . \qquad \text{N3.3.2}$$

From the definition of ξ (equation 3.2.12), we have

$$\frac{d\xi}{dr_o} = \frac{r_o d(\delta r)/dr - \delta r}{r_o^2} \quad \text{or}, \quad d\xi = \frac{d(\delta r)}{r_o} - \frac{\delta r \, dr_o}{r_o^2} \qquad \text{N3.2.3}$$

Eliminating $d(\delta r)$ from (N3.3.2) with the aid of (N3.3.3), we have again to the first order

$$\int_o^{r'} r_o^2 \delta\rho \, dr_o = - \int_o^{r'} \rho_o r_o^2 (3 + r_o \frac{d\xi}{dr_o}) \, dr_o \ . \qquad \text{N3.3.4}$$

Equation (N3.3.4) must hold for all values of r'. This can only be true if the integrands are identically equal. Thus,

$$\frac{\delta\rho}{\rho_o} = - (3\xi + r_o \frac{d\xi}{dr_o}) \qquad \text{N3.3.5}$$

3.4   In an attempt to simplify equation (3.2.15), let us consider the second integral.

$$3 \int_o^M \frac{P_o}{\rho_o} (\gamma-1) \, r_o \frac{d\xi_o}{dr_o} e^{i\sigma t} \, dm(r) = 3 \int_o^{R_o} P_o(\gamma-1) r_o \frac{d\xi_o}{dr_o} e^{i\sigma t} 4\pi r_o^2 \, dr_o \qquad \text{N3.4.1}$$

Integrating by parts we obtain

$$12\pi e^{i\sigma t} \int_0^R \left[\frac{d\xi_o}{dr_o}\right] [P_o(\gamma - 1) r_o^3] dr_o$$

$$= 12\pi e^{i\sigma t} P_o(\gamma - 1) r_o^3 \xi_o \Big|_0^{R_o} \quad \text{N3.4.2}$$

$$- 12\pi e^{i\sigma t} \int_0^R \xi_o \frac{d}{dr_o} [P_o(\gamma - 1) r_o^3] dr_o .$$

At this point we shall impose the boundary conditions that $P_o \to 0$ at $r_o = R_o$ and $\xi_o \to 0$ at $r_o = 0$. The first of these conditions is the familiar condition of stellar structure which essentially defines the surface of the gas sphere. The second condition is required by assuming that the radial pulsation is a continuous function. With these conditions, the integrated part of (N3.4.2) vanishes and the remaining integral becomes

$$\int_0^R \xi_o \frac{d}{dr_o}[P_o(\gamma-1) r_o^3] dr_o = \int_0^R P_o \xi_o r_o^3 dr + \int_0^{R_o}(\gamma-1)\xi_o \frac{dP_o}{dr_o} r_o^3 dr_o$$

$$+ 3\int_0^R P_o(\gamma-1)\xi_o r_o^2 dr_o . \quad \text{N3.4.3}$$

Conservation of momentum (i.e., hydrostatic equilibrium) requires that

$$\frac{dP_o}{dr_o} = - \frac{Gm(r_o)\rho_o}{r_o^2} . \quad \text{N3.4.4}$$

Therefore,

$$4\pi r_o^3 \frac{dP_o}{dr_o} = - \frac{4\pi G r_o^2 \rho m(r)}{r_o} = - \frac{Gm(r_o)}{r_o} \frac{dm(r_o)}{dr_o} = \frac{d\Omega_o}{dr_o} . \quad \text{N3.4.5}$$

Making use of the second form of (N3.4.5) to simplify the second integral in (N3.4.3), and the definition of dm(r) to simplify the other two integrals in (N3.4.3), we obtain the following expression for the variation of the kinetic energy (3.2.15).

$$2\delta T = -9 e^{i\sigma t} \int_0^M \frac{P_o \xi_o}{\rho_o}(\gamma-1) dm(r_o) + 3 e^{i\sigma t} \int_0^M \frac{P_o \xi_o r_o}{\rho_o} \frac{d\gamma}{dr_o} dm(r_o)$$

$$+ 3 e^{i\sigma t} \int_0^{M_o} \xi_o(\gamma-1) d\Omega_o + 9 e^{i\sigma t} \int_0^M \frac{P_o \xi_o}{\rho_o}(\gamma-1) dm(r_o) . \quad \text{N3.4.6}$$

The above expression is obtained by making use of the previously mentioned definition and substituting it into (N3.4.3), then into (N3.4.2), and finally into (3.2.15). Since the first and last integrals are identical except for the difference in sign, they vanish from the expression and

$$2\delta T = 3 e^{i\sigma t} \int_0^M \frac{P_o \xi_o}{\rho_o} \frac{d}{dr_o} dm(r) + 3 e^{i\sigma t} \int_0^{\Omega_o} \xi_o(\gamma-1) d\Omega_o . \quad \text{N3.4.7}$$

3.5 Let us define

$$\eta_i = \delta x_i \quad \text{N3.5.1}$$

Now the definition of the mass within a given volume in cartesian coordinates becomes

$$m(V) = \iiint \rho dx_1 dx_2 dx_3 . \quad \text{N3.5.2}$$

The conservation of mass requires, as it did in Section 2, that the variation of the mass is zero. Thus, taking the variations of (N3.5.2), we obtain

$$\delta m(V) = 0 = \iiint \delta\rho\, dx_1 dx_2 dx_3 + \iiint \rho\, d(\delta x_1) dx_2 dx_3$$
$$+ \iiint \rho\, dx_1 d(\delta x_2) dx_3 + \iiint \rho\, dx_1 dx_2 d(\delta x_3) . \qquad \text{N3.5.3}$$

Re-writing the last three integrals, we have

$$\iiint \delta\rho\, dx_1 dx_2 dx_3 = - \iiint \rho \sum_{i=1}^{3} \frac{d(\delta x_i)}{dx_i}\, dx_1 dx_2 dx_3 . \qquad \text{N3.5.4}$$

Since $\delta x_i = \eta_i$, we have by the chain rule

$$\frac{d(\delta x_i)}{dx_i} = \frac{d\eta_i}{dx_i} = \sum_{j=1}^{3} \frac{dx_j}{dx_i} \frac{\partial \eta_i}{\partial x_j} . \qquad \text{N3.5.5}$$

However, since the x_i's are linearly independent and the η_i's are just the variation of these coordinates, not only is the second term in the product (under the summation sign of N3.5.5) zero if $i = j$ but, so is the first term.

Thus, as might be expected from the orthogonality of the x_i's we have

$$\frac{d\eta_i}{dx_i} = \frac{\partial \eta_i}{\partial x_i} . \qquad \text{N3.5.6}$$

Substitution of this into (N3.5.4), yields

$$\iiint \delta\rho\, dx_1 dx_2 dx_3 = - \iiint \rho \sum_{i=1}^{3} \frac{\partial \eta_i}{\partial x_i}\, dx_1 dx_2 dx_3 . \qquad \text{N3.5.7}$$

Equation (N3.5.7) must hold for any volume of integration. This can only be true if the integrands themselves are equal. Thus, we finally obtain

$$\frac{\delta\rho}{\rho} = - \sum_{i=1}^{3} \frac{\partial \eta_i}{\partial x_i} . \qquad \text{N3.5.8}$$

3.6 Suppose the displacement **n** takes place with the slow continuous movement so that

$$\widetilde{\boldsymbol{u}} = \frac{d\boldsymbol{n}}{dt} . \qquad \text{N3.6.1}$$

Then, if the electrical conductivity of the medium is infinite, the *time variation* of the electric field is

$$\Delta \mathbf{E} = - \widetilde{\boldsymbol{u}} \times \mathbf{H} . \qquad \text{N3.6.2}$$

However, Maxwell's equations for an infinitely conducting medium require that

$$\nabla \times (\Delta \mathbf{E}) = - \frac{\partial}{\partial t} \Delta \mathbf{H} . \qquad \text{N3.6.3}$$

Combining (N3.6.2) and (N3.6.3), we have

$$- \frac{\partial (\Delta \mathbf{H})}{\partial t} = \nabla \times (-\widetilde{\boldsymbol{u}} \times \mathbf{H}) . \qquad \text{N3.6.4}$$

The integral form of (N3.6.4) is just

$$\Delta \mathbf{H} = \nabla \times (\boldsymbol{n} \times \mathbf{H}) . \qquad \text{N3.6.5}$$

But, from the definition of the time and space variations and the equation relating to the total and partial time derivatives, we know that

$$\Delta \mathbf{H} = \frac{\partial \mathbf{H}}{\partial t} dt \qquad \text{N3.6.6}$$

and

$$\frac{\partial}{\partial t} = \frac{d}{dt} - \mathbf{v} \cdot \nabla , \qquad \text{N3.6.7}$$

100

which results in

$$\Delta \mathbf{H} = [\frac{d\mathbf{H}}{dt} - \sum_j \frac{dx_i}{dt} \cdot \nabla \mathbf{H}] \, dt \, . \qquad \text{N3.6.8}$$

However,

$$\frac{d}{dt} = \sum_i \frac{\partial}{\partial x_i} \frac{dx_i}{dt} \, , \qquad \text{N3.6.9}$$

While definition of the space variation $\delta \mathbf{H}$ is

$$\delta \mathbf{H} = \sum_i \frac{\partial \mathbf{H}}{\partial x_i} \, dx_i \, . \qquad \text{N3.6.10}$$

Combining (N3.6.10), (N3.6.9), and (N3.6.8), we have

$$\Delta \mathbf{H} = \delta \mathbf{H} - \sum_i \frac{dx_i}{dt} \, dt \cdot \nabla \mathbf{H} \, . \qquad \text{N3.6.11}$$

Noting that the variation of a linearly independent quantity may be interpreted as the total differential of the quantity,

$$dx_i = \sum_i \frac{\partial x_i}{\partial x_j} \, dx_j = \delta x_i \qquad \text{N3.6.12}$$

we obtain

$$\Delta \mathbf{H} = \delta \mathbf{H} - \boldsymbol{n} \cdot \nabla \mathbf{H} \qquad \text{N3.6.13}$$

Combining (N3.6.13) and (N3.6.5), we finally obtain an equation for the variation of the magnetic field ($\delta \mathbf{H}$).

$$\delta \mathbf{H} = \nabla \times (\boldsymbol{n} \times \mathbf{H}) + (\boldsymbol{n} \cdot \nabla) \mathbf{H} \qquad \text{N3.6.14}$$

Now the curl of a cross-product may be written as

$$\nabla \times (\boldsymbol{n} \times \mathbf{H}) = (\mathbf{H} \cdot \nabla) \boldsymbol{n} - \mathbf{H}(\nabla \cdot \boldsymbol{n}) - (\boldsymbol{n} \cdot \nabla)\mathbf{H} + \boldsymbol{n}(\nabla \cdot \mathbf{H}) \, . \qquad \text{N3.6.15}$$

Since the divergence of $\mathbf{H}$ is always zero, the last term vanishes. Combining this expression for the curl of a cross-product with (N3.6.14) yields

$$\delta \mathbf{H} = (\mathbf{H} \cdot \nabla) \boldsymbol{n} - \mathbf{H}(\nabla \cdot \boldsymbol{n}) - (\boldsymbol{n} \cdot \nabla)\mathbf{H} + (\boldsymbol{n} \cdot \nabla)\mathbf{H} \qquad \text{N3.6.16}$$

The last two terms are identical except for opposite signs and thus cancel out. The remaining expression may be written in component form as follows:

$$\delta H_i = \sum_j [H_j \frac{\partial n_i}{\partial x_j} - H_i \frac{\partial n_j}{\partial x_j}] \, . \qquad \text{N3.6.17}$$

3.7 In the section 3.2 we went through an extensive argument (see Note 3.6, Equation N3.6.16), to show that

$$\delta \mathbf{H} = (\mathbf{H} \cdot \nabla) \, \delta \mathbf{r} - \mathbf{H}(\nabla \cdot \delta \mathbf{r}) \, , \qquad \text{N3.7.1}$$

where δr plays the role of η in that discussion. The easiest way to evaluate the first term is in cartesian coordinates, remembering that ξ_o is constant. Then,

$$(\mathbf{H} \cdot \nabla) \delta \mathbf{r} = \sum_i H_i \frac{\partial (\xi x_i)}{\partial x_i} = \xi \mathbf{H} \, . \qquad \text{N3.7.2}$$

The second term can be evaluated the same way, so that

$$\mathbf{H}(\nabla \cdot \delta \mathbf{r}) = H_j \sum_i \frac{\partial (\xi x_i)}{\partial x_i} = 3\xi \mathbf{H} \qquad \text{N3.7.3}$$

So, the variation of the magnetic field has taken the particularly simple form

$$\delta \mathbf{H} = -2\xi \, \mathbf{H}_o \qquad \text{N3.7.4}$$

References

1. Ledoux, P. (1945), Ap. J. 102, 143.
   _____. (1958), Handbuch der Physik, Springer-Verlag, Berlin, Gottingen, Heidelberg, pp. 605-687.
2. Chandrasekhar, S. (1963), Ap. J. 137, 1185.
3. Chandrasekhar, S. and Fermi, E. (1953), Ap. J. 118, p. 116.
4. Ledoux, P. and Walraven, (1958), Handbook der Physik, Vol. 51, pp. 431-592, Springer-Verlag, Berlin, Gottingen, Heidelberg.
5. Chandrasekhar, S. and Limber, D. N. (1959), Ap. J. 119, p. 10.
6. Chandrasekhar, S. (1961), Hydrodynamic and Hydromagnetic Stability, Oxford University Press, London, p.
7. _____. (1939), Stellar Structure, Dover Pub., Inc. (1957), p. 53.
8. Hunter, C. (1977), Ap. J. 213, p.497.
9. Lyden-Bel, D. and Ostriker, J. P. (1967), M.N.R.A.S. 136, p. 293.
10. Tassoul, J. L. and Ostriker, J. P. (1968), Ap. J. 154, p. 613.
11. Ostriker, J. P. and Tassoul, J. L. (1969), Ap. J. 155, p. 987.
12. Ostriker, J. P. and Bodenheimer (1973), Ap. J. 180, p. 171.
13. Ostriker, J. P. and Peebles, P. J. E. (1973), Ap. J. 186, 467.
14. Bardeen, J. M., Friedman, J. L., Schutz, B. F. and Sarkin, R. (1977), Ap. J. Lett. 217, p. L49.
15. Milne, E. A. (1925), Phil. Mag. S. 6, Vol. 50, p. 409-419.

4. Some Applications of the Virial Theorem

1. Pulsational Stability of White Dwarfs

By now, I hope the reader has been impressed by the wide range of problems which can be dealt with by the virial theorem. Some of the problems mentioned in the last chapter indicate the type of insight which can be achieved through use of virial theorem, however, the type of objects which were explicitly discussed, notably normal stars, are currently judged by the naive to be well understood. In order to illustrate the power of this remarkable theorem, I cannot resist discussing some objects about which even less is known. We shall see that at least one commonly held tenant of stellar structure, while leading to nearly the correct numerical result, is conceptually wrong.

During the 1960's advances in observational astronomy presented problems requiring theoreticans to postulate the existence of a wide range of objects previously considered only of academic interest. These terms, like supermassive stars, neutron stars, and black holes became 'household' words in the literature of astrophysics. Many of the objects were clearly so condensed as to require the application of the general theory or some other gravitational theory for their description. When one postulates the existence of a "new" object it is always wise to subject that object to a stability analysis. This is particularly important for highly collapsed objects as the time scale for development of the instability will be very short. Since general relativistic effects can usually be viewed as an effective increase in the gravitational force, one would expect its presence to decrease the stability of objects in which it is important. What came as a surprise is the importance of these effects where one would normally presume them to be of little or no importance.

Apparently inspired by a comment of R. P. Feymann in 1963, W. A. Fowler noted that effects of general relativity would lead to previously unexpected instabilities in supermassive stars[1]. Noting that the conditions for this instability also exist in massive white dwarfs, Chandrasekhar and Tooper[2] showed by means of rather detailed calculations that a white dwarf would become unstable when its radius shrank to about 246 Schwarzschild radii or on the order of 1000 km. This corresponds to a mass about 1.5% below the well-known Chandrasekhar limiting mass for degenerate objects. During the next 15 years, this instability received a great deal of attention and I will not attempt to fully recount it here. Rather, let us examine with the aid of hindsight and the virial theorem, how this result could be anticipated without the need of detailed calculations.

One can see that the stability analysis coupled with the post-Newtonian form of the virial theorem given in Chapter 2 (eq. 2.4.15) would serve as the basis for investigating this effect. However the estimation or calculation of the relativistic terms on the right hand side of 2.4.15 is extremely difficult. Instead, by assuming spherical symmetry we may start with the spherically symmetric equation of motion given by Meltzer and Thorne[3] as did Fowler[4] and follows the formalism of Chapter 3. Thus

$$y \frac{d}{dt}(y dr/dt) = -\frac{1}{\rho}\frac{dP}{dr}\left[\frac{1 + \frac{y^2}{c^2}(\frac{dr}{dt})^2 - 2Gm(r)/rc^2}{1 + P/\rho c^2}\right]$$
$$- \frac{Gm(r)}{r^2} - 4\pi G \rho r/c^2 \qquad 4.1.1$$

where

$$y = (\rho + P/c^2)/\rho_o .$$

If we confine our attention to objects nearly in equilibrium, no large scale radial motions can exist, thus the term involving $(dr/dt)^2$ can be neglected and 4.1.1 becomes

$$y^2 \frac{d^2r}{dt^2} = -\frac{1}{\rho}\frac{dP}{dr}\left[\frac{1 - 2Gm(r)/rc^2}{1 + P/\rho c^2}\right] - \frac{Gm(r)}{r^2} - \frac{4\pi GPr}{c^2}, \quad 4.1.2$$

or in the post-Newtonian approximation (i.e., keeping only terms of the order $1/c^2$)

$$\rho y^2 \frac{d^2r}{dt^2} = -\frac{dP}{dr}\left[1 - \frac{P}{\rho c^2} - \frac{2Gm(r)}{rc^2} - \cdots - \right]$$

$$= \frac{Gm(r)\rho}{r^2} - \frac{4\pi GP\rho r}{c^2}. \quad 4.1.3$$

In hydrostatic equilibrium, dP/dr is given by Oppenheimer-Volkoff as

$$\frac{dP}{dr} = -\frac{G(\rho + P/c^2)[m(r) + 4\pi r^3 P/c^2]}{r^2(1 - 2Gm(r)/rc^2)}, \quad 4.1.4$$

or

$$\frac{dP}{dt} \approx -\frac{G\rho m(r)}{r^2} + \frac{1}{c^2}\left[\frac{2G^2\rho m^2(r)}{r^4} + \frac{G\rho m(r)}{r^2}\right.$$

$$\left. + 4\pi r G\rho^2\right] + O(1/c^4). \quad 4.1.5$$

If we retain dP/dr explicitly for the expansion of the first term of 4.1.3 all other relativistic terms of 4.1.5 will be of the order $1/c^4$ in the product in 4.1.3. Thus, 4.1.3 becomes

$$\rho y^2 \frac{d^2r}{dt^2} = -\frac{dP}{dr} - \frac{Gm(r)\rho}{r^2}\left[1 + \frac{P}{\rho c^2} + \frac{2Gm(r)}{rc^2} + \cdots + \right]$$

$$- \frac{4\pi GrP\rho}{c^2}. \quad 4.1.6$$

Now if we again form Lagrange's identity by multiplying by r and integrating over all volume, we get

$$\int_V (y^2 \rho r \frac{d^2 r}{dt^2}) dV = - \int_0^R 4\pi r^3 dP - \int_V \frac{Gm(r)\rho}{r} dV - \int_V \frac{Gm(r)P}{rc^2} dV$$

$$- 2 \int_V (\frac{Gm(r)}{rc^2})^2 dV - 4\pi \int_V \frac{GPr^2 \rho}{c^2} dV . \qquad 4.1.7$$

The last integral can be integrated by parts,[4.1] so that

$$\int_V \frac{4\pi GPr^2 \rho}{c^2} dV = \int_V \frac{G^2 m^2(r)\rho}{r^2 c^2} dV . \qquad 4.1.8$$

With somewhat less effort the first integral becomes

$$\int_0^R 4\pi r^3 dP = 4\pi r^3 P \Big|_0^R - \int_0^R 12\pi r^2 P dr = - 3 \int_V P dV . \qquad 4.1.9$$

Putting the results of 4.1.8 and 4.1.9 into 4.1.7 and rewriting the left hand side in terms of a relativistic moment of inertia we get

$$\frac{1}{2} \frac{d^2 I_r}{dt^2} = 3 \int_V P dV - \Omega - \frac{1}{c^2} \int_V \frac{Gm(r)P}{r} dV$$

$$- \frac{3}{c^2} \int_V \frac{G^2 m^2(r)\rho}{r^2} dV \qquad 4.1.10$$

which is equivalent to 2.4.15 of Chapter 2 for spherical stars but vastly simpler. Although this approach, which is basically due to Fowler[4], lacks the rigor of the EIH post-Newtonian approach, it does yield the same results for spherical stars nearly in hydrostatic equilibrium. It is worth noting that the relativistic correction terms are of the same mixed energy integrals as those that appear in equation 2.4.15.

Taking the variation of 4.1.10 as we did in Chapter 3, we have

$$\frac{1}{2} \frac{d^2}{dt^2}(\delta I_r) = 3\delta \left[\int_V P dV\right] - \delta\Omega - \frac{1}{c^2} \delta\left[\int_V \frac{Gm(r)PdV}{r}\right] \qquad 4.1.11$$

$$- \frac{3}{c^2} \delta\left[\int_M \frac{G^2 m^2(r) dm(r)}{r^2}\right]$$

As before, let us suppose that the variation of these quantities results from a variation of the independent variable δr. Further assume that $\delta r/r = \xi_0 e^{i\sigma t}$ where ξ_0 is constant and the variation is adiabatic. Since we can write the internal heat energy density as $(\Gamma_1 - 1)u = P$, the first term becomes

$$3 \delta \int_V P dV = 3 \langle \Gamma_1 - 1 \rangle \delta \mathcal{U} . \qquad 4.1.12$$

Equation 3.2.10 (Chapter 3) leads to

$$\delta \Omega = - \xi \Omega . \qquad 4.1.13$$

The variation of the relativistic correction terms can be computed as follows: Let

$$\Omega_1 = \frac{3}{2} \int_o^M \frac{G^2 m^2(r)}{r^2 c^2} dm(r) \qquad 4.1.14$$

so that the variation of the last term in 4.1.11 is

$$2\delta\Omega_1 = \frac{3}{c^2} \int_o^M G^2 m^2(r) \delta \left(\frac{1}{r^2} \right) dm(r) = - 2\xi\Omega_1 . \qquad 4.1.15$$

It is convenient (particularly for the relativistic terms) to normalize by the dimensionless quantity $(2GM/Rc^2)$. Thus,

$$\Omega_1 = \frac{3}{2} Mc^2 \int \frac{1}{4} \left(\frac{2GM}{Rc^2} \right)^2 \left(\frac{m^2(r)}{M} \right) \left(\frac{R}{r} \right)^2 \frac{dm(r)}{M}$$

$$= \frac{3}{8} Mc^2 \left[\frac{2GM}{Rc^2} \right]^2 \int_o^1 (q/x)^2 dq \qquad 4.1.16$$

where the dimensionless variables are $q = (m(r)/M)$, and $x = r/R$.

The remaining terms in 4.1.11 can be normalized in a similar way by making use of the homologous dependance of P. That is

$$P = \eta G m^2(r)/r^4 \qquad 4.1.17$$

where η is a dimensionless scale factor. Thus

$$P_1 = \frac{1}{c^2} \int_V \frac{Gm(r)PdV}{r^2} = \frac{1}{c^2} \int_0^R \frac{4\pi G^2 m^3(r) r^2 dr}{r^5}$$

$$= Mc^2 \left[\frac{2GM}{Rc^2}\right]^2 \int_0^1 \pi\eta\left(\frac{q}{x}\right)^3 dx \quad .$$

4.1.18

As in 4.1.16, the integral in 4.1.18 is dimensionless and determined by the equilibrium model. Thus the remaining equation is

$$\delta P_1 = -2\xi P_1 \quad .$$

4.1.19

replacing P with $u(\Gamma_1 - 1)$ as with the first term and letting

$$\mathcal{U}_1 = \frac{1}{c^2} \int_V \frac{u Gm(r)}{r} dV = \frac{1}{\langle\Gamma_1 - 1\rangle} P_1$$

4.1.20

4.1.11 becomes

$$\frac{1}{2} \frac{d^2}{dt^2}\left(\delta I_r\right) = 3 \langle\Gamma_1 - 1\rangle \delta\mathcal{U} - \delta\Omega$$

$$+ \langle\Gamma_1 - 1\rangle \delta\mathcal{U}_1 - 2\delta\Omega_1 \quad .$$

4.1.21

Now, since the internal energy $\mathcal{U}$ is coupled with all other terms including the relativity terms we shall eliminate it in a somewhat different fashion than in Chapter 3. Since the total energy must be constant, its variation is zero. Thus

$$\delta E = 0 = \delta\mathcal{U} - \delta\Omega + \delta\mathcal{U}_1 - \delta\Omega_1$$

4.1.22

and 4.1.21 becomes

$$\frac{1}{2} \frac{d^2}{dt^2}\left(\delta I_r\right) = \langle 3\Gamma_1 - 4\rangle \delta\Omega - 2\langle\Gamma_1 - 1\rangle \delta\mathcal{U}_1$$

$$+ \langle 3\Gamma_1 - 5\rangle \delta\Omega_1 \quad .$$

4.1.23

Substituting in the variations from (4.1.13, 4.1.15, and 4.1.19) into 4.1.23 and noting that two time differentiations of the perturbation will give a σ^2 in the first term, equation 4.1.23 becomes

$$\sigma^2 I_r \bar{y} = <3\Gamma_1-4> \Omega_o - 4<\Gamma_1-1> \mathcal{U}_1 + 2 <3\Gamma_1-5>\Omega_1 \quad . \quad 4.1.24$$

Making one last normalization of Ω_o which for polytropes is

$$\Omega_o = \left(\frac{3}{5-n}\right) \frac{GM^2}{R} = \frac{1}{(5-n)} \frac{3}{2} \left(\frac{2GM}{Rc^2}\right) Mc^2 \quad . \quad 4.1.25$$

and calling the dimensionless integrals in 4.1.16 and 4.1.18, ζ_1 and ζ_2 respectively, equation 4.1.24 becomes

$$\sigma^2 I_r \bar{y} = Mc^2 \left(\frac{2GM}{Rc^2}\right) \left\{ \frac{3}{2(5-n)} <3\Gamma_1-4> \right.$$
$$\left. - \left(\frac{2GM}{Rc^2}\right) (4\zeta_1 <\Gamma_1-1> + 2\zeta_2 <5-3\Gamma_1>) \right\} \quad . \quad 4.1.26$$

Since the average relativity factor $\bar{y}$ is always positive, this expression can be used as a stability criterior as in Chapter 3. That is

$$\left(\frac{R_s}{R_o}\right) (4\zeta_1 <\Gamma_1-1> + 2\zeta_2 <5-3\Gamma_1>) < \left[\frac{3<3\Gamma_1-4>}{2(5-n)}\right] \quad 4.1.27$$

where we have used the fact that the Schwarzschild radius is just $2GM/c^2$. We can now use this to investigate the stability of white dwarfs as they approach the Chandrasekhar limiting mass. As this happens, the equation of state approaches that of relativistically degenerate electron gas and the internal structure, that of a polytrope of index 3. As $n \to 3$, $\Gamma_1 \to 4/3$, and the system becomes unstable. Thus let

$$\Gamma_1 = 4/3 + \varepsilon \quad\quad\quad 4.1.28$$

and 4.1.27 becomes (using Fowler's[4] values for ζ_1 and ζ_2)

$$\frac{9\varepsilon}{4} - \frac{R_s}{R_o}(\frac{4}{3}\zeta_1 + 2\zeta_2) = \frac{9}{4}\varepsilon - 2.535\frac{R_s}{R_o} \geqslant 0 \qquad 4.1.29$$

or $\qquad \frac{R_o}{R_s} > 1.127/\varepsilon$.

Thus, if you imagine a sequence of white dwarfs of increasing mass, the value of (R_o/R_s) will monotonically decrease as a result of the mass radius relation for white dwarfs and $(1/\varepsilon)$ will monotonically increase as the configuration approaches complete relativistic degeneracy. Clearly, the point must come where the system becomes unstable and collapses. However, in order to find that point, we need an estimate of how ε changes with increasing mass. For that we turn to an interesting paper by Faulkner and Gribben[5] who show[4.2]

$$\varepsilon \cong \frac{2x^{-2}}{3} \qquad 4.1.30$$

where x is the Chandrasekhar degeneracy parameter.
So, our instability condition can be written as

$$\frac{R_o}{R_s} > 1.69 \, \overline{x^2} \, . \qquad 4.1.31$$

All that remains is to estimate an average value of the degeneracy parameter x which we can expect to be much larger than 1. Remember from Chandrasekhar[6]

$$\rho_e = B \, x^3 = (8\pi m_e^4 c^3/3h)x^3 \, . \qquad 4.1.32$$

Now neglecting inverse β decay the local density will be roughly given by $\rho = m_p \rho_e/m_e$ and

$$x^3 = [3h/(8\pi m_e^3 c^3 m_p)]\rho \, . \qquad 4.1.33$$

Let ρ be given by its average value so that

$$\overline{x^2} = [9h/32\pi m_e^3 c^3 m_p]^{2/3} \frac{M^{2/3}}{R^2} \, . \qquad 4.1.34$$

110

Normalizing R by Schwarzschild radius we get

$$\overline{x^2} = 7.02 \times 10^6 \, (M_\odot/M)^{4/3} \, (R_s/R_\odot)^2 . \qquad 4.1.35$$

This can be rigorously combined with the mass radius relation for white dwarfs and equation 4.1.31 to provide value for (R_o/R_s). However, since this entire argument is illustrative it will suffice to assume that the mass is roughly the limiting mass for white dwarfs. Then 4.1.31 becomes

$$\left(\frac{R_o}{R_s}\right) > 228 \left(\frac{M_\odot}{M}\right)^{4/9} \approx 210 \qquad 4.1.36$$

which is in remarkable agreement with the more precise figure of Chandrasekhar and Tooper[2] of 246. It is most likely that the discrepancy arises from the rather casual way of estimating $\overline{x^2}$ since it will be affected by both the type of volume averaging to determine $\overline{\rho}$ and the details of the equation of state used in relating ρ to ρ_e.

2. The Influence of Rotation and Magnetic Fields on White Dwarf Gravitational Instability

At this point the reader is likely to complain that the derivation indicating the presence of an instability resulting from general relativity has been anything but brief. The length results largely from a somewhat different approach to the general relativistic term than used earlier. That the approach succeeds at all is largely a result of presumed spherical symmetry. However, to further demonstrate the effacy of this approach let us consider what impact rotation and magnetic fields may have on the results of the last section. That this is indeed a reasonable question I need only point out that Fowler found that a very small amount of rotation would stabilize larger supermassive stars against the gravitational instability. However, the situation for white dwarfs is quite different. Here the gravitational field is proportionally much stronger with γ being driven to 4/3 by the equation of state and not the radiation field. Thus we may expect that a much larger rotational energy field is required to bring about stability than is the case for supermassive stars. In spite of this expectation, we shall assume that the effects of rotation and magnetic fields are not so extreme as to significantly alter the spherical symmetry.

Under these conditions, the Newtonian approach of Chapter 3 will suffice to calculate the terms to be added to the equations of motion and to perform the required variational analysis. In Chapter 3, we defined the rotational kinetic energy $\mathcal{T}_3$ and magnetic energy $\mathcal{M}$ as

$$\mathcal{T}_3 = \int_0^{\mathcal{L}} \tfrac{1}{2} \, \omega d\mathcal{L} \, ,$$

$$\mathcal{M} = \int_V \frac{H^2}{8\pi} \, dV \quad \quad 4.2.1$$

which have variations

$$\delta \mathcal{T}_3 = -2\xi \mathcal{T}_3(0) \quad \quad 4.2.2$$

and

$$\delta \mathcal{M} = -\xi \mathcal{M}_o \, .$$

Adding this to the variational form of Lagrange's identity in Section 1 (eq. 4.1.21), we get

$$\tfrac{1}{2} \frac{d^2}{dt^2} (\delta I_r) = 3 \langle \Gamma_1 - 1 \rangle \delta \mathcal{U} - \delta \Omega + 2 \delta \mathcal{T}_3 + \delta \mathcal{M} \quad \quad 4.2.3$$

$$+ \langle \Gamma_1 - 1 \rangle \, \delta \mathcal{U}_1 - 2 \delta \Omega \, .$$

and now the condition on the variation of the total energy becomes

$$\delta E = 0 = \delta \mathcal{U} - \delta \Omega + \delta \mathcal{T}_3 + \delta \mathcal{M} + \delta \mathcal{U}_1 - \delta \Omega_1 \quad \quad 4.2.4$$

which enables us to re-write 4.2.4 as

$$\tfrac{1}{2} \frac{d^2}{dt^2} (\delta I_r) = \langle 3\Gamma_1 - 4 \rangle (\delta \Omega - \delta \mathcal{M}) - 2 \langle \Gamma_1 - 1 \rangle \delta \mathcal{U}_1 \quad \quad 4.2.5$$

$$+ \langle 5 - 3\Gamma_1 \rangle (\delta \mathcal{T}_3 - \delta \Omega_1) \, .$$

Substituting in the values for the variations we get an expression analogous to 4.1.24

$$\sigma^2 I_r \, \bar{y} = \langle 3\Gamma_1 - 4 \rangle (\Omega_o - \mathcal{M}_o) - 4 \langle \Gamma_1 - 1 \rangle \mathcal{U}_1 \quad \quad 4.2.6$$

$$+ 2 \langle 3\Gamma_1 - 5 \rangle (\Omega_1 - \mathcal{T}_3(0)) \, .$$

These expressions differ from those in Chapter 3 only because the gravitational potential energy is taken here to be positive.

In order for us to proceed further it will be necessary to normalize both the rotational energy and magnetic field by something. Let us consider the case for ridged rotation so that

$$\mathcal{T}_3 = \frac{1}{2}\omega^2 I_z = \frac{1}{3}\omega^2 I. \qquad 4.2.7$$

Here we are ignoring the relativistic corrections to I and take $I = \alpha MR^2$. In addition, let us normalize the angular velocity ω by the critical value for a Roche model. Then

$$\omega^2 = w^2(8\ GM/27R_o^3) = \frac{4\ w^2 c^2}{27\ R_o^2}\left(\frac{2GM}{R_o c^2}\right) \qquad 4.2.8$$

This certainly does not imply that we are in any way assuming that white dwarfs are represented by a Roche model but rather that it merely provides us with a convenient scale factor. Thus

$$\mathcal{T}_3 = \frac{4\alpha}{81} w^2\ Mc^2\left(\frac{2GM}{R_o c^2}\right). \qquad 4.2.9$$

In a similar manner let us normalize the magnetic energy $\mathcal{M}_o$ by the energy sufficient to bring about dissruption of the star. In Chapter 3 we showed that if other effects were absent then $\mathcal{M}_o > |\Omega_o|$ would disrupt the star. Using this as the normalization constant we have

$$\mathcal{M}_o = \eta^2\left[\frac{3}{2(5-n)}\right] Mc^2 \left(\frac{2GM}{R_o c^2}\right) \qquad 4.2.10$$

Under these conditions we can expect the maximum values for w and η to be

$$\left.\begin{array}{c} w < 1 \\ \eta \ll 1 \end{array}\right\}, \qquad 4.2.11$$

and in any event the assumption of sphericity will probably break down for $w > 0.8$ and $\eta > 0.3$.

Putting these values for $\mathcal{T}_3$ and $\mathcal{M}$ along with the previously determined values for Ω, $\mathcal{U}_1$, and Ω_1, into

4.2.6 we can arrive at stability conditions analogous to 4.1.28. Namely

$$\left[\frac{R_s}{R_o}\right] \left(4\zeta_1 <\Gamma_1 -1> + 2\zeta_2 <5-3\Gamma_1>\right) < \frac{3<3\Gamma_1-4>(1-\varkappa^2)}{2(5-n)} + \frac{8\alpha w^2}{81}$$

As before, let us pass to the case where n = 3, so that

$$\left[\frac{R_s}{R_o}\right] \left(\frac{4}{3}\zeta_1 + 2\zeta_2\right) < \frac{9}{4}\varepsilon(1-\varkappa^2) + 8\alpha w^2/81 . \qquad 4.2.13$$

With $\alpha = 0.113$ for polytropes of n = 3 and again using Fowler's[4] values for ζ_1 and ζ_2 this becomes

$$\left(\frac{R_s}{R_o}\right) < 0.8876\, \varepsilon(1-\varkappa^2) + 4.4 \times 10^{-3} w^2 . \qquad 4.2.14$$

Using the same analysis for ε as before

$$\left(\frac{R_s}{R_o}\right) < 8.4 \times 10^{-8} (1-\varkappa^2)\left(\frac{M}{M_o}\right)^{4/3}\left[\frac{R_o}{R_s}\right]^2 + 4.4 \times 10^{-3} w^2 \qquad 4.2.15$$

and taking M to be near the Chandrasekhar limit, we have

$$\left(\frac{R_o}{R_s}\right)^3 + \left(\frac{R_o}{R_s}\right) \frac{4.096 \times 10^4 w^2}{(1-\varkappa^2)} - \frac{9.3 \times 10^{-6}}{(1-\varkappa^2)} > 0 . \qquad 4.2.16$$

For a point of reference it is worth mentioning the size of the normalization quantities so that various values of w and $\varkappa$ may be held in perspective. If we assume that we are dealing with objects on the order of 10^3 km, then the disruption field is of the order of 3×10^{15} gauss while the critical equatorial velocity would be about 10^4 km/sec. The largest observed fields in white dwarfs reach 10^8 gauss and although it can be argued that larger fields may be encountered in more massive white dwarfs, fields in excess of 10^{12} gauss would not seem to be supported by observations. Thus, a plausible upper value of $\varkappa$ would be on the order of 10^{-3}.

If, for the moment we neglect rotation 4.2.14 can be written as

$$\left(\frac{R_o}{R_s}\right) > 210\ (1-\mathcal{H}^2)^{-1/3}\ . \qquad 4.2.17$$

So it is clear that the only effect the field has is to act with general relativity to further de-stabilize the star. However, for the field to make any appreciable difference it will have to be truly large. For our plausible upper limit of $\mathcal{H} \sim 10^{-3}$ the effect is to increase the radius at which the instability sets in by about 1%.

The situation regarding rotation is slightly more difficult to deal with as the resulting inequality is a cubic. With $\mathcal{H} = 0$, 4.2.14 becomes

$$\left(\frac{R_o}{R_s}\right)^3 + \left(\frac{R_o}{R_s}\right)\ 4.096 \times 10^4 w^2 - 9.3036 \times 10^{-6} > 0\ . \qquad 4.2.18$$

An additionally useful expression for the equatorial velocity which corresponds to a given w is

$$V_{eq.} = R\omega = \frac{2c}{3\sqrt{3}}w\left(\frac{R_s}{R_o}\right)^{1/2} = 1.15 \times 10^5\ w\left(\frac{R_s}{R_o}\right)^{1/2}\ \text{km/sec}. \qquad 4.2.19$$

If we pick a few representative values of w and solve 4.2.17 by means of the general cubic we get the results below for the associated values of (R_o/R_s) and V_{eq}.

Rotational Effects on the
White Dwarf Instability Limit

w	0	0.1	0.5	1.0
V_{eq} (km/sec)	0	794	4128	9453
R_o/R_s	210.3	209.6	194	148

Even these few values are sufficient to indicate that although rotation helps to stabilize the star in the sense of allowing it to attain a smaller radius before collapse, nevertheless, like magnetic fields, the effect is small. One may choose to object to the assumption of rigid rotation as being too conservative. However, it is clear from the development that for either rotation or magnetic fields to really play an important role the total energy stored by either mechanism must approach that in the gravitational field. In order to do this with differential rotation, the differential velocity field would have to be alarmingly high. It seems likely that the resulting shear would produce significant dynamical instabilities.

Thus, we have seen that neither magnetic fields nor rotation can significantly alter the fact that a white dwarf will become unstable at or about 1000 km. Classically the star reaches this point when it is within, but less than, a few percent of the Chandrasekhar limiting mass. So it is not the limiting mass resulting from the change in the equation of state that keeps us from observing more massive white dwarfs. Rather it is the presence of general relativistic instability that destroys any more massive objects.

It is quite simple to dismiss this argument as 'nit-picking' as the mass at which the instability occurs is nearly identical to the Chandrasekhar limiting mass. However, when one tries to generalize the results of one problem to another, it is conceptual errors such as this that may lead to much more serious errors in the generalization. As we shall see in the next section, this is indeed the case with neutron stars.

3. Stability of Neutron Stars

A second class of objects whose existence became well established during the 1960's are the neutron stars. It is a commonly held misconception that a neutron star is nothing more than a somewhat collapsed white dwarf, since the masses are thought to be similar. In reality the ratios of typical white dwarf radii to neutron stars is suspected to be nearly 1000 which is just about the ratio of the sun's radius to that of a typical white dwarf. A similar misconception relates to the notion of the neutron star limiting mass. It is popular to suggest that since the mass limit in white dwarfs arises as a result of the change in the equation of state so a similar change in the equation of state for a neutron star yields a limiting mass for these objects. It is true that a limiting mass exists for neutron stars but this limit does not necessarily arise from a change in the equation of state.

Let us consider a very simple argument to dramatize this point. The equation of the state change that results in the Chandrasekhar limit occurs because the electrons achieve relativistic velocities. If this were to happen in neutron stars, the configuration would still have to satisfy the virial theorem.

For the moment let us ignore the effects of general relativity and just consider the special relativistic virial theorem as we derived in Chapter 2 (equation 2.3.9),

$$\frac{1}{2} \frac{d^2}{dt^2} I_r = \Omega + T + \int_V \beta \tau dV \qquad 4.3.1$$

where $\beta = (1 - v^2/c^2)^{1/2}$.

As the neutrons become relativistic $\beta \to 0$ and $T = \alpha Mc^2$ where $\alpha \gg 1$. We may write the gravitational potential energy as

$$\Omega = -\eta \frac{GM^2}{R_o} = -\frac{\eta Mc^2}{2} (R_s/R_o) . \qquad 4.3.2$$

The variational form of the virial theorem will require that

$$T + \Omega < 0 \qquad 4.3.3$$

so that

$$Mc^2 [\alpha - (\eta/2)(R_s/R_o)] < 0 \qquad 4.3.4$$

or

$$\left(\frac{R_o}{R_s}\right) > \eta/2\alpha .$$

Since η is of the order of unity and $\alpha \gg 1$, this would require that the object have a radius less than the Schwarzschild radius in order to be stable against radial pulsations. This is really equivalent to invoking Jacobi's stability condition on the total energy.

This simplistic argument can be criticized on the grounds that it ignores general relativity which can be viewed as increasing the efficiency of gravity. Perhaps the "increased gravity" would help stabilize the star against the rapidly increasing internal energy. This is indeed the case for awhile. However, based on the analysis in Section 1, as the value of Γ approaches 4/3 the same type of instability which brought about the collapse of the white dwarfs will occur in the neutron stars. The exact value of R_o/R_s for which this happens will depend on the exact nature of the equation of state as well as details of model construction. However, since the general relativistic correction terms will be much larger than in the case of white dwarfs, we should expect the value of Γ to depart farther from the relativistic limit of 4/3 than before. That this is indeed the case is clearly shown by Tooper[7] in considering the general properties of relativistic adiabatic fluid spheres. He concludes that the instability always occurs before the gas has become relativistic at high pressures.

Unfortunately we cannot quantitatively apply the results of Section 1 since the way in which Γ approaches 4/3 (more properly the way in which it departs from 5/3) depends in detail on the equation of state. However, we may derive some feeling for the way in which the instability sets in by assuming the compression has driven the value of Γ down from 5/3 to 3/2 (i.e., just half way to its relativistic value. Substitution of $\Gamma = 3/2$ in equation 4.1.27 and using Fowler[4] values of ζ_1 and ζ_2 for a polytrope of index 2 gives a stability limit of

$$\left(\frac{R_o}{R_s}\right) > 4.3 \quad . \qquad 4.3.5$$

Thus R_o for a neutron star would have to be greater than about 12 km. Since typical model radii are of the order of 10 km, 3/2 is probably a representative value of Γ, yet it is still far from the relativistic value of 4/3. This argument further emphasizes the fact that it is the general relativistic instability which places an upper limit on the size of the configuration, not the equation of state becoming relativistic.

The discussion in Section 2 would lead us to believe that neither rotation nor magnetic fields can seriously modify the onset of the general relativistic instability. This can be made somewhat quantitative by evaluating equation 4.2.10 for a polytrope of index $n = 2$. However, in order to do this, we must re-evaluate the moment of inertia weighting factor α. A crude estimate here will suffice since we are neglecting an increase of perhaps a factor of 2 due to general relativistic terms. One can show by integrating $4\pi \int_o^R r^2 \rho dr$ by parts in Emden polytropic variables that

$$\alpha = [1 + (\frac{6}{\xi_1^2} \int_o^\xi \xi\Theta d\xi) / (\xi_1^2 \frac{d\Theta}{d\xi}\big|_{\xi_1})] \qquad 4.3.6$$

which for n = 2 very approximately gives α = 0.345. Substitution into 4.2.11 then yields

$$\left(\frac{R_s}{R_o}\right) < 0.234 \ (1-\eta^2) + 3.2 \times 10^{-2} w^2 . \qquad 4.3.7$$

The effects of magnetic fields and rotation are qualitatively the same for neutron stars as for white dwarfs. However, as R_o is only a few times the Schwarzschild radius, the normalizing fields and rotational velocities are truly immense. For a neutron star with a 10 km radius the magnetic field corresponding to $\eta = 1$ in of the order of 10^{18} gauss while the rotational velocity corresponding to w = 1 would be about 10% the velocity of light. These values vastly exceed those for the most extreme pulsar. Thus, barring modification to the equation of state resulting from these effects, they can largely be ignored in investigating neutron star stability. This result is exactly in accord with what one might have expected on the basis of the white dwarf analysis.

We began this discussion by indicating that the analysis would be very simplistic and yet we have attained some very useful qualitative results. Since the mass-radius law for any degenerate equation of state (excepting small technical wiggles) will provide for stars whose radius decreases with increasing mass, we can guarantee that the resulting decrease in Γ will give rise to an unstable configuration at a few Schwarzschild radii. Thus, there will exist an upper limit to the mass allowable for a neutron star. The origin of this limit is conceptually identical to that for white dwarfs. Furthermore, as for white dwarfs, this limit can be modified by the presence of magnetic fields and rotation only for the most extreme values of each.

One may argue that in discussing effects of general relativity we have included terms of O $(1/C^2)$ and that higher order effects may be important. While this is true regarding such items as gravitational radiation, none of these terms should be important unless the configuration becomes smaller than 2-3 Schwarzschild radii. Even then, they are unlikely to affect the qualitative behavior of the results.

Very little has been said about the large volume of work relating to the equation of state for neutron degenerate matter. This is most certainly not to deny its existence, just its relevance.

One of the strong points of this approach is that insight can be gained into the global behavior of the object without undue concern regarding the microphysics. This type of analysis is a probe intended to ascertain what effects are important in the construction of a detailed model and what may be safely ignored. It cannot hope to provide the information of a detailed structural model but only point the way toward successful model construction.

4. Additional Topics and Final Thoughts

It would be possible and perhaps even tempting to continue demonstrating the efficacy of the virial theorem in stellar astrophysics almost indefinitely. However, attempting to exhaust the possible applications of the virial theorem is like trying to exhaust the applicability of the conservation of momentum. I would be remiss if I did not indicate at least some other possible areas in which the virial theorem can lend insight.

In Chapter 3, Section 4 we discussed the variational effect of the surface terms resulting from the application of the divergence theorem. These terms are generally neglected and for good reason. In most cases the bounding surface can be chosen so as to include the entire configuration. In instances where this is not the case, such as with magnetic fields, the term is still generally negligible. If one considers the form of the surface terms given by 2.5.19 (Chapter 2), compared to the volume contribution (equation 2.5.18), the ratio of the scalar value is

$$\mathcal{A} = \frac{\int_S P_o \mathbf{r} \cdot d\mathbf{s}}{\int_V P dV} \simeq \frac{P_o}{<P>} \qquad 4.4.1$$

where $<P>$ is the average value of the internal pressure. Since, in any equilibrium configuration the pressure must be a monotone increasing function as one moves into the configuration, $\mathcal{A} < 1$. In general it is very much less than one. However, in the case where $\gamma \to 4/3$ the variational contribution of the surface pressure approaches $3P_o \int \mathbf{r} \cdot d\mathbf{s}$, while the internal pressure and Newtonian gravity contributions vanish. Such terms are then available to combine with the effects of general relativity. Since they are of the same sign as the general relativistic terms the surface terms will only serve to increase the instability of the entire configuration. This results in an increase of the radius at which white dwarfs become unstable. Thus, the accretion of matter onto the surface of such objects may cause them to collapse sooner than one might otherwise expect.

Since magnetic fields also are usually assumed to increase inward, the influence of magnetic surface terms will be similar to that of a non-zero surface pressure. Like the surface pressure terms, they will in general be small compared to the internal contributions. Only in case of a system with an effective γ approaching $4/3$ could these small terms be expected to exert a trigger effect on the resulting configuration.

There is one instance in which the use of the surface terms can be significant. One of the most attractive aspects of this entire approach is that it can deal with the properties of an entire system. Indeed the spatial moments are taken in order to achieve that result. However it is interesting to consider the effects of applying the virial theorem to a sub volume of a larger configuration. Clearly, as one shrinks the volume to zero he recovers the equations of motion themselves multiplied by the local positional coordinate. If one considers a case intermediate to these limits and investigates the stability of a sub volume which could include the surface of the star, it would be possible to analyze the outer layer for instabilities which might not be globally apparent. It is true that a local stability criterion would be sufficient to locate such instabilities but one could not be sure how the instabilities would propagate without carrying out a large structural analysis. This latter effect can be avoided by utilizing a sub global form of the virial theorem. Under these conditions one might expect the surface terms to be the dominant terms of the resulting expression.

In discussing some of the more bizzare and contemporary aspects of stellar structure it is easy to overlook the role played by the virial theorem in the development of the classical theory of stellar structure. It is the virial theorem which provides the theoretical basis for the definition of the Kelvin-Helmholtz contraction time. This is just the time required for a star to radiate away the available gravitational potential energy at its present luminosity. It is the virial theorem which essentially tells how much of the gravitational energy is available. Thus if the contraction liberating the potential energy is uniform and d^2I/dt^2 is zero then the total kinetic energy always must be

$$T = -1/2\, \Omega \quad . \qquad 4.4.2$$

This makes the other half of the gravitational energy available to be radiated away. The Kelvin-Helmholtz contraction time for polytropes is thus

$$T = 3GM^2/2(5-n)RL$$
$$\simeq \frac{4.5 \times 10^7}{(5-n)} \left(\frac{M}{M_\odot}\right)^2 \left(\frac{R_\odot\, L_\odot}{R\, L}\right) \text{ years} \quad . \qquad 4.4.3$$

Reasoning that this provided an upper limit to the age of the sun Lord Kelvin challenged the Darwinian theory of evolution on the sound ground that 13 million years (i.e., KHT for a polytrope with $\gamma = 5/3$) was not long enough to allow for the evolutionary development of the great diversity of life on the planet. The reasoning was flawless, only the initial assumption that the sun derived its energy from gravitational contraction which was plausible at the time, failed to withstand the development of stellar astrophysics.

Another aspect of classical stellar evolution theory is clarified by application of the virial theorem. All basic courses in astronomy describe post-main sequence evolution by pointing out that the contraction of the core is accompanied by an expansion of the outer envelope. Most students find it baffling as to why this should happen and are usually supplied with unsatisfactory answers such as "it's obvious" or "it's the result of detailed model calculations" which freely translated means "the computer tells me it is so." However, if the virial theorem is invoked, then once again any internal re-arrangement of material that fails to produce sizable accelerative changes in the moment of inertia will require that

$$2T + \Omega = 2E - \Omega = 0 . \qquad 4.4.4$$

Since the only way that the star can change its total energy E without outside intervention is by radiating it away to space, any internal changes in the mass distribution which take place on a time scale less than the Kelvin Helmholtz contraction time will have to keep the total energy and hence the gravitational potential energy constant. Now

$$\Omega = - \alpha M^2/R . \qquad 4.4.5$$

where α is a measure of the central condensation of the object, so, as the core contracts and α increases, R will have to increase in order to keep Ω constant. In general the evolutionary changes in a star do take place on a time scale rather less than the contraction time and thus we would expect a general expansion of the outer layer to accompany the contraction of the core. The microphysics which couple the core contraction to the envelope expansion is indeed difficult and requires a great deal of computation to describe it in detail. However the mass distribution of the star places constraints on the overall shape it may take on during rapid evolution processes. It is in the understanding of such global problems that the virial theorem is particularly useful.

    I have attempted throughout this book to emphasize that these global properties are the very essence of the virial theorem. The centrality of taking spatial moments of the equations of motion to the entire development of the theorem demonstrates this with more clarity than any other aspect. Although this global structure provides certain problems when the development is applied to continuum mechanics nothing is encountered within the framework of Newtonian mechanics which is insurmountable. Only within the context of general relativity may there lie fundamental problems with the definition of spatial moments. Even here the first order theory approximation to general relativity yields an unambiguous form of the virial theorem. In addition, certain specific time independent or at least slowly varying cases of the non-approximated equations also yield unique results. Thus one can realistically hope that a general formulation of the virial theorem can be made although he must expect that the interpretation of the resultant space-time moments will not be intuitivity obvious.

The rather recent development of the virial theorem provides us with a dramatic example of the fact that theories do not develop in an intellectual vacuum. Rather they are pushed and shoved into shape by the passage of the time. Thus we have seen the virial theorem born in an effort to clarify thermodynamics and arising in parallel form in classical dynamics. However the similarity did not become apparent until the implications of the ergodic theorem inspired by statistical mechanics were understood. Although sparsely used by the early investigators of stellar structure, the virial theorem did not really attract attention until 1945 when the global analysis aspect provided a simple way to begin to understand stellar pulsation. The attendant stability analysis implied by this approach became the main motivation for further development of the tensor and relativistic forms and provides the primary area of activity today. Only recently has the similarity of virial theorem development to that of other conservation laws been clearly expounded.

Recent criticism of some work utilizing the virial theorem, incorrectly attacks the theorem itself as opposed to analyzing the application of the theorem and the attendant assumptions. This is equivalent to attacking a conservation law and serves no useful purpose. Indeed it may by rhetorical intimidation turn some less sophisticated investigators aside from from consideration of the theorem in their own problems. This would be a most unfortunate result as by now even the most skeptical reader must be impressed by the power of the virial theorem to provide insight into problems of great complexity. Although there is a trade-off in that a complete dynamical description of the system is not obtainable, certain general aspects of the system are analyzable. Even though some might claim a little knowledge to be a dangerous thing, I prefer to believe that a little knowledge is better than none at all. Thus, the perceptive student of science will utilize the virial theorem to provide a 'first look' at problems to see which are of interest. Used well this first look will not be his last.

Through the course of this book we have examined the origin of the virial theorem, noted its development and applicability to a wide range of astrophysical problems, and it is irresistable to contemplate briefly its future growth. In my youth the course of future events always seemed depressingly clear but turned out to be generally wrong. Now, in spite of a better time base on which to peer forward, the new future seems at best "seen through a glass darkly", and I am mindful that astronomers have not had an exemplary record as predictors of future events.* Nevertheless, there may be one or two areas of growth for the virial theorem on which we can count with some certainty.

Immediate problems which seem ideally suited to the application of the virial theorem certainly include exploration into the nature of the energy source in QSO's. Perhaps one will finally observe that the gravitational energy of assembly of a galaxy or its components is of the same order as the estimated energy liberated by a QSO during its lifetime. The virial theorem implies that half of this energy may be radiated away. Thus, it would appear that one need not look for the source of such energy but rather be concerned with the details of the "generator".

---

* It is said that the great American astronomer, Simon Newcomb "proved" that heavier than air flight was impossible and that after the Wright Brothers flew, it was rumored that he maintained it would never be practical as no more than two people could be carried by such means.

Perhaps future development will consider applications of the virial theorem as represented by Lagrange's identity. To date the virial theorem has been applied to systems in or near equilibrium. It is worth remembering that perhaps the most important aspect of the theorem is that it is a global theorem. Thus systems in a state of rapid dynamic change are still subject to its time dependent form.

Lately, as a consequence of discovering that the universe is not a quiet place, theoreticians became greatly excited about the properties of objects undergoing unrestrained gravitational collapse. It is logical to suppose that sooner or later they will become interested in the effects of such a collapse upon fields other than gravitation (i.e., magnetic or rotational), that may be present. The virial theorem provides a clear statement on how the energy in such a system will be shifted from one form to another as soon as one has determined d^2I/dt^2. Future investigation in this area may be relevant to phenomena ranging from novae to quasars.

Perhaps the most exciting and at the same time least clear and speculative development in which the virial theorem may play a role involves its relationship to general relativity. This is a time of great activity and anticipatory excitement in fundamental physics and general relativity in particular. Perhaps through the efforts of Stephan Hawking and others, and as Denis Sciama has noted, we are on the brink of the unification of general relativity, quantum mechanics, and thermodynamics. Thermodynamics is the handmaiden of statistical mechanics and it is here through the application of the ergodic theorem that the virial theorem may play its most important future role.

You may remember that in Chapter 2, difficulty in the interpretation of moments taken over spacetime frustrated a general development of the virial theorem in general relativity and it was necessary to invoke first order approximations to the relativistic field equations. In addition the ergodic theorem seems inexorably tied to the nature of reversible and inreversible processes. The advances in relating general relativity to thermodynamics bring these areas and theorems into direct conceptual confrontation and may perhaps provide the foundations for the proper understanding of time itself.

Notes to Chapter 4

4.1 The last integral can be integrated by parts[4.1], so that

$$\int_V \frac{4\pi G P r^2 \rho}{c^2} dV = \frac{4\pi G P r^2 m(r)}{c^2} \Big|_0^R - \int_0^R \frac{4\pi r^2 G m(r)}{c^2} \frac{dP}{dr} dr$$

$$- \int_0^R \frac{8\pi G m(r) P r dr}{c^2}$$

$$= \int_V \frac{G^2 m^2(r) \rho}{r^2 c^2} dV$$

$$- \frac{8\pi G}{c^2} m(r) P r \Big|_0^R + \frac{8\pi G}{c^2} \int_0^R m(r)[P + (dP/dr)] dr$$

$$= \int_0^R \frac{G^2 m^2(r) \rho}{r^2 c^2} dV + \frac{8\pi G}{c^2} m(r) P(r) \Big|_0^R$$

$$- \frac{8\pi G}{c^2} \int_0^R m(r) \frac{dP}{dr} dr + \frac{8\pi G}{c^2} \int_0^R m(r) \frac{dP}{dr} dr \qquad \text{N4.1.1}$$

$$\int_V \frac{4\pi G P r^2 \rho}{c^2} dV = \int_V \frac{G^2 m^2(r) \rho}{r^2 c^2} dV \quad . \qquad \text{N4.1.2}$$

4.2 Starting with the polytropic equation of state

$$P = K\rho^\gamma \qquad \text{N4.2.1}$$

it is not hard to convince yourself that

$$\gamma = \frac{d\ln P}{d\ln \rho} = \frac{\rho}{P} \frac{dP}{d\rho} \qquad \text{N4.2.2}$$

This can be reduced to a single parameter by considering Chandrasekhar's parametric equation of state for a nearly relativistic degenerate gas[6]

$$P = A f(x) \qquad \rho = B x^3 \qquad \text{N4.2.3}$$

where $f(x) = x(2x^2-3)(x^2+1)^{1/2} + 3 \sinh^{-1}(x)$.
Now consider the behavior of $f(x)$ as $x \to \infty$.

$$f(x) \simeq x(2x^2-3)(x + \tfrac{1}{2} x^{-1} + \ldots +) = 2x^4 + x^2 - 3x^2 - \tfrac{3}{2} x^{-1} + \ldots + \qquad \text{N4.2.4}$$

or

$$f(x) \simeq 2(x^4 - x^2) \qquad \text{N4.2.5}$$

Similarly

$$\frac{dP}{d\rho} = \frac{dP}{dx} \frac{dx}{d\rho} = \frac{A}{B} \frac{(8x^3 - 4x)}{3 x^2} \quad . \qquad \text{N4.2.6}$$

Thus

$$\varepsilon = \gamma - 4/3 \simeq \frac{B x^3}{2A(x^4-x^2)} \left[\frac{A}{B} \frac{(8 x^3 - 4x)}{3x^2}\right]^{-4/3} = \frac{2}{3}\left[\frac{(2x^2-1)}{x^2-1}\right] - 4/3 \qquad \text{N4.2.7}$$

or

$$\varepsilon = \tfrac{2}{3}\left[1 + \frac{x^2}{x^2-1}\right] - 4/3 = \tfrac{2}{3}[1 + x^{-2} + x^{-4} + \ldots +)] - 4/3 = \frac{2x^{-2}}{3} \quad . \qquad \text{N4.2.8}$$

References

1. Thorne, K. S. (1972), Stellar Evolution, ed. Hang-Yee Chin and Amador Muriel MIT Press, Cambridge, Mass. p. 616.
2. Chandrasekhar, S. and Tooper, R. F. (1964), Ap. J. 130, p. 1396.
3. Meltzer, D. W. and Thorne, K. S. (1966), Ap. J. 145, p. 514.
4. Fowler, W. A. (1966), Ap. J. 144, p. 180.
5. Faulkner, J. and Gribbin, J. R. (1966), Nature, Vol. 218, p. 734-7.
6. Chandrasekhar, S. (1957), An Introduction to the Study of Stellar Structure, Dover Pub., p. 360-361.
7. Tooper, R. F. (1965), Ap. J. 142, p. 1541.

Symbol Definitions and First Usage

Since this work contains a large number of concepts symbolically expressed, I felt it might be useful if a brief definition of these symbols existed in some place for purposes of reference. In general, the use of bold face type denotes a vector quantity, while a style of type known as 'engraver's text' is used for tensor-like quantities of rank 2 or higher. Subscripted Old English type is used to represent the components of these tensors. Various other type faces have been employed to provide symbolic representation of scalar quantities which appear throughout this work. What follows is a list of the meaning of these symbols and where they first appear.

Scalars and Special Parameters

Symbol	Meaning	First Used
A	An arbitrary scalar	2.4.10
A	Constant in the degenerate equation of state	N4.2.3
B	Constant in the degenerate equation of state	N4.2.3
D	The magnitude of the electric displacement vector	N2.4.1
E	Total energy of a system	1.1.8
F	The magnitude of the radiative flux	1.1.10
G	Gravitational constant	1.1.7
G	A temporary quantity	1.2.2
H	The magnitude of the magnetic field intensity	N2.4.1
I	Moment of inertia about a coordinate origin	1.2.3
I_r	Moment of inertia including relativistic effects	2.3.15
I_z	Moment of inertia about the Z axis	3.2.21
J	Trace of the Maxwell tensor	2.4.3
K	The constant of proportionality in the polytropic equation of state	N4.2.1
L(r)	Stellar luminosity	1.1.10
L_Θ	Solar luminosity	4.4.3
$\mathscr{L}$	The magnitude of the angular momentum vector	3.3.19
M	Stellar or configuration total mass	3.2.4
M_Θ	Solar mass	3.5.22
$\mathfrak{M}$	Total internal magnetic energy	2.5.18
N	The particle number density	2.5.21
P	Total scalar pressure	1.1.6
P_g	Total gas pressure	N3.2.7
P_1	A relativistic correction term	4.1.18
Q	Arbitrary macroscopic system parameter	1.5.1
<Q>	Average Q	1.5.1
Q	Arbitrary point-defined system property	2.2.3
Q_p	Surface pressure term Defined in	3.4.12 / 3.4.8
Q_m	Magnetic surface term Defined in	3.4.11 / 3.4.3
Q_i	The generalized forces acting on a system	3.5.1
R	The Rydberg constant	3.3.2
R	Stellar or configuration radius	3.2.20
R_s	The Schwarzschild radius	4.1.27
R_Θ	The solar radius	3.5.22
$\mathfrak{R}$	Total rotational kinetic energy	2.5.17
S	The 'creation rate' or collision term of the Boltzmann transport equation	1.1.1
$\overline{S}$	The velocity averaged 'creation rate'	1.1.2

129

Scalars and Special Parameters (Continued)

Symbol	Meaning	First Used	Symbol	Meaning	First Used
S	The surface enclosing a volume V	2.5.12	k	Boltzmann's constant	2.5.21
T	Total kinetic energy of a system	1.2.5	m_i	Mass of the ith particle	1.2.1
$\overline{T}$	The time averaged kinetic energy of a system	1.2.15	m_e	The electron mass	4.1.32
$<T>$	The phase averaged kinetic energy of a system	3.5.8	m_p	The proton mass	4.1.33
T_o	A period of time	1.2.14	$m(r)$	The mass of a spherical system interior to r	1.1.10
T	The Kelvin-Helmholtz contraction time	4.4.3	$m(V)$	The mass of a spherical system within a volume V	N3.5.2
$\mathcal{T}$	The pulsation period of a star	3.2.24	n	Exponent of the attractive force law	1.2.8
$\mathfrak{T}$	The kinetic or gas temperatures	2.5.20	n	Polytropic index	4.1.25
$\mathfrak{T}_1$	The kinetic energy of mass radial gas motions	3.3.6	q	A dimensionless mass	4.1.16
$\mathfrak{T}_2$	The kinetic energy associated with thermal motions	3.3.2	q_i	The ith linearly independent system coordinate	3.5.1
$\mathfrak{T}_3$	The kinetic energy associated with macroscopic rotational motions	3.3.6	r	Magnitude of the radial coordinate	1.1.10
$\mathfrak{U}$	The total potential energy of a system	1.2.12	r_i	Radial coordinate of the ith particle	1.2.2
$\overline{\mathfrak{U}}$	The time averaged potential energy of a system	1.2.15	r_{ij}	Separation between the ith and jth particles	1.2.8
$\mathfrak{U}_1$	The post-Newtonian correction times c^2 to the Newtonian internal energy	4.1.20	s	Proper length	2.3.3
$\mathfrak{U}$	The total internal energy due to heat	2.5.20	t	Time	1.1.1
V	A volume usually enclosing a system	1.4.1	t_o	An initial time	1.5.1
W	A general-relativistic super-potential 'energy'	2.4.13	u	Magnitude of the stream velocity vector	2.4.6
Z	A general-relativistic super-potential 'energy'	2.4.13	u	Internal thermal energy density	4.1.20
a_{ij}	The force law proportionality constant	1.2.8	v	Magnitude of the velocity vector	1.1.3
c	The speed of light	2.3.3	w	Fractional angular velocity (in units of the Roche critical velocity)	4.2.8
c_p	Specific heat of constant pressure	2.5.21			
c_v	Specific heat at constant volume	2.5.21	x_α	Components of Minkowski space	2.3.3
h	Planck's constant	4.1.32	x	The parametric variable in the degenerate equation of state	4.1.30
$\varkappa$	The dimensionless magnetic field intensity normalized by the disruption field	4.2.10	x	A dimensionless length	4.1.16
			x	A cartesian coordinate	3.3.18
			y	A cartesian coordinate	3.3.18
			y	The normalized relativistic density	4.1.1

Scalars and Special Parameters (Concluded)

Symbol	Meaning	First Used	Symbol	Meaning	First Used
Γ_1	The first adiabatic constant of Chandrasekhar	4.1.31	ζ_1	A dimensionless parameter of polytropic structure	4.1.26
Π	The internal energy of a relativistic gas times c^2	2.4.7	ζ_2	A dimensionless parameter of polytropic structure	4.1.26
ϕ	An arbitrary potential	1.2.8	η	A dimensionless scale factor	4.1.17
$\langle\Phi\rangle$	A relativistic super potential	2.4.13	η	A proportionality constant	4.3.2
$\varPhi$	A relativistic super potential	2.4.18	θ	The polar angle in a spherical coordinate system	3.3.18
χ	Local energy generated from viscous or non-conservative forces	1.1.8	Θ	The polytropic temperature	4.3.6
X	A higher order super potential	2.2.10	ξ	The normalized Lagrangian variation of the radial coordinate	3.2.12
$\mathcal{X}$	A higher order super potential	2.2.10	ξ	The polytropic radial coordinate or Emden variable	4.3.6
φ	A relativistic super potential	2.4.13	ξ_1	The polytropic Emden radius	4.3.6
ψ	The Newtonian potential	1.1.4	ρ	The local matter density	1.1.2
$\mathcal{A}$	The ratio of surface to average pressure	4.4.1	ρ_e	The electric charge density	2.5.4
Ω	The Newtonian gravitational potential energy	1.2.13	ρ_e	The electric (mass) density	4.1.32
$\overline{\Omega}$	The time averaged Newtonian potential energy	1.2.16	ρ^*	Modified matter-energy density	2.4.6
$\langle\Omega\rangle$	The phase averaged Newtonian potential energy	3.5.8	σ	Modified matter-energy density	2.4.7
Ω_1	The post-Newtonian correction to the Newtonian potential energy times c^2	4.1.14	σ	Pulsational frequency	3.2.12
α	A parameter measuring the degree of mass concentration	4.2.8	τ	"Relativistic" kinetic energy density	N2.3.5
β_i	A proportionality constant for velocity dependent forces	1.3.3	ϕ	The source of the relativistic super potential	2.4.9
β	$(1 - v^2/c^2)^{1/2}$	4.3.1	ϕ	Azimuthal polar coordinate	3.4.5
β	The angle between the local H field and the radius vector	3.4.14	ψ	Density of points in phase-space	1.1.1
γ	$(1 - v^2/c^2)^{1/2}$	2.3.3	ω	Magnitude of the angular velocity	N2.6.4 3.3.18
γ	The ratio of specific heats	2.5.23			
ϵ	The energy generation rate from non-vicuous sources	1.1.9			
ϵ	The total energy density	2.6.3			
ϵ	The potential energy density	N2.3.6			
ϵ	A perturbation parameter	4.1.28			
ϑ	Thermal kinetic energy density	2.5.21			

Vectors and Vector Components

Symbol	Meaning	First Used
$\mathbf{A}$	An arbitrary vector	N2.4.9
$\mathbf{B}$	The magnetic field vector	2.5.3
B_i	The components of $\mathbf{B}$	2.5.2
$\mathbf{D}$	The electric displacement vector	2.5.3
D_i	The components of $\mathbf{D}$	2.5.2
$\mathbf{E}$	The electric field vector	2.5.3
E_i	The components of $\mathbf{E}$	2.5.2
$\mathbf{F}$	The radiative flux vector	1.1.8
$\mathbf{F}_i$	The total force acting on the ith particle	Page 2
$\mathbf{F}_{ij}$	The conservative force vector between the ith and jth particles	1.2.7
$\mathbf{G}$	An arbitrary vector	N2.4.9
$\mathbf{H}$	The magnetic field intensity vector	2.5.3
$\hat{\mathbf{H}}_o$	A unit vector along $\mathbf{H}$	3.4.14
H_i	The cartesian components of the magnetic field intensity vector	2.5.2
$\mathbf{K}$	The relativistic linear momentum density in the post-Newtonian approximation	2.4.11
$\mathbf{Y}$	A general relativistic super potential	2.4.8
$\mathbf{f}$	The local force vector	1.1.1
$\mathbf{f}_i$	The force vector acting on the ith particle	1.2.1
$\boldsymbol{f}$	The local force density vector	1.4.3
$\boldsymbol{f}_f$	The frictional force density	2.5.24
$\boldsymbol{l}$	The net local angular momentum density	2.5.16
$\mathbf{p}_i$	The momentum vector of a particle	1.2.1
P_i	The components of the momentum vector of a particle	1.1.1
$\boldsymbol{p}$	The local momentum density vector	1.4.3
$\mathbf{r}_i$	The radius vector of the ith particle	1.2.1
r_i	The components of the radius vector	2.5.23
$\hat{\mathbf{r}}$	The unit vector in the direction of $\mathbf{r}$	Page 2, 3.3.39
$\mathbf{s}$	The vector "creation" rate	1.1.4
$d\mathbf{s}$	The differential surface normal vector	2.3.2
$\mathbf{u}$	The local stream velocity	1.1.2
u_i	The components of stream velocity	2.1.4
u_α	The components of the 4-velocity	2.3.3
$\tilde{\mathbf{u}}$	Lagrangian displacement velocity field vector	N3.6.1
$\mathbf{v}$	The local velocity vector	1.1.1
$\mathbf{v}_i$	The velocity vector of the ith particle	1.2.1
$\mathbf{w}$	The local peculiar velocity in a rotating or non-inertial coordinate frame	2.5.6
$\overline{\mathbf{w}}$	Velocity dependent force vector	1.3.1
$\mathbf{x}$	A cartesian coordinate vector	2.6.3
x_i	The components of the cartesian coordinate vector	1.1.1
$\mathbf{z}_i$	A velocity independent force vector of the ith particle	1.3.1
$\boldsymbol{n}$	The Lagrangian displacement vector	3.3.38
n_i	The cartesian components of displacement vector	N3.5.1
$\boldsymbol{\omega}$	The angular velocity field vector	2.5.5
ω_i	The cartesian components of field vector	N2.5.4
$\tilde{\boldsymbol{\omega}}$	The local residual angular velocity field in a rotating coordinate frame	N2.6.2
$\tilde{\omega}_i$	The cartesian components of the local angular velocity field	2.5.15

Tensors and Tensor Components

Symbol	Meaning	First Used
$\mathfrak{F}$	The "frictional" tensor providing the source of frictional forces	2.5.24
$\mathfrak{I}$	The moment of inertia tensor	2.1.5
$\mathfrak{I}_{ij}$	The components of moment of inertia tensor $\mathfrak{I}$	2.1.11
$\mathfrak{J}$	The Maxwell stress-energy tensor or energy momentum tensor	2.3.1
$\mathfrak{J}_{ij}$	The components of the Maxwell tensor $\mathfrak{J}$	2.3.4
$\mathfrak{L}$	The volume angular momentum tensor	N2.6.3, 2.5.10
$\mathfrak{L}_{ijk}$	The components of $\mathfrak{L}$	2.5.15
$\mathfrak{M}$	The magnetic energy tensor	2.5.14
$\mathfrak{M}_{ij}$	The components of magnetic energy tensor $\mathfrak{M}$	
$\mathfrak{P}$	The pressure tensor	1.1.4
$\mathfrak{P}_g$	The gas pressure tensor	2.5.1
$\mathfrak{R}_{ij}$	The components of the Ricci tensor	2.4.3
$\mathfrak{S}$	The surface energy tensor	2.5.12
$\mathfrak{S}_{ij}$	The components of energy tensor $\mathfrak{S}$	2.5.12
$\mathfrak{T}$	The kinetic energy tensor	2.1.5
$\mathfrak{T}_{ij}$	The components of energy tensor $\mathfrak{T}$	2.1.11
$\mathfrak{U}$	The potential energy tensor	2.1.5
$\mathfrak{U}_{ij}$	The components of the energy tensor	N2.2.1
$\mathbf{1}$	The unit tensor	2.2.7
δ_{ij}	The components of the unit tensor or Kronecker's "delta"	2.4.5
ϵ_{ijk}	The components of the Levi-Civita tensor	N2.5.4
h_{ij}	The components of the metric perturbation tensor	2.4.1
g_{ij}	The components of the metric tensor	2.4.1

Index

Angular Momentum, 59, 73
   conservation, 30, 37, 72
Angular Velocity, 48
   critical, 113
   local, 59
Arnold, V. I., 21
Average Energy
   kinetic, 2
   potential, 2
Average stream velocity, 9
Averages:
   phases, 19-22, 86
   time, 14, 19-22
   velocity, 9
Avez, A., 21
Birkhoff's Theorem, 21
Black holes, 103
Boltzmann Transport Equation, 5,9,28
Boltzmann, L., 19
Carnot, N. L. S., 1
Center of Mass
   acceleration, 32
Chandrasekhar limit, 104,109,114,116
Chandrasekhar, S., 2,12,27,31,33,39,63,
   77,78,79,104,110,111
Charge density, 18
Claussius, R. J. E., 1
Conservation laws, 11, 35, 46
Conservation of:
   angular momentum, 30, 37, 94
   energy, 11
   linear momentum, 44
   mass, 9, 16, 31, 36, 64, 98
   momentum, 99
Coriolis force, 49, 51, 59, 93
Creation rate, 9
Critical angular velocity, 113
Density
   charge, 18
   electron, 110
   force, 16
   kinetic energy, 17
   matter, 9, 16, 41
   momentum, 16
   relativistic matter, 41
EIH approximation, 39, 46, 106
Einstein field equations, 40
Einstein, Infeld, Hoffman
   approximation, (see EIH)
Electron density, 110
Electrostriction, 47
Energy
   all forms (see specific forms)
   conservation, 11
Equations of motion:
   Newtonian, 12
   perturbed, 62
   relativistic, 42
   for a zero resistivity gas, 78

Equatorial velocity, 115
Ergodic hypothesis, 20
Euclidean metric, 39
Euler-Lagrange Equations, 10, 31, 40 with
   magnetic fields, 47
Farquhar, I. E., 19, 21
Faulkner, J., 110
Fermi, E., 3, 27, 63, 92
Feynmann, R. P., 104
Force,
   Coriolis,49, 51, 59, 93
   friction, 15, 53
   generalized, 83
   Lorentz, 16, 48
Force density, 16
Fowler, W. A., 104,106,110,111,118
Friction forces, 15, 53
Generalized forces, 83
Goldstein, H., 12,48
Gravitational potential, 14
Gravitational potential energy, 34, 97,
   113, 123
   variation, 65
Gribben, J. R., 110
Hawking, S. 126
Heat energy, 52, (see also Internal energy)
Hunter, C., 93
Hydrostatic equilibrium, 10, 105,
   (see also Conservation of momentum)
Instability, (see Stability)
Internal energy, 34,45, 52, 107
   variation, 69-70
Jacobi's stability criterion, 86, 117
Jacobi, K., 3
Jeans' stability criterion, 86
Jeans, Sir J., 1
Kelvin-Helmholtz Contraction Time, 122
Kelvin, Lord, 122
Kinetic energy, 2,12,17,34,47,69,97
   average, 2
   pulsational, 69
   relativistic, 58, 117
   rotational, 69
   thermal, 69
   variation, 65
Kinetic energy density, 17
Kurth, R., 11
Lagrange, J. L., 3
Lagrange's indentity, 3,11,14,18,44,49
   51,54,69,87,105
   special relativistic form, 34
Landau, L. D., 11, 34
Lebovitz, N., 27, 31
Ledoux, P., 2.63.95
Lewis' Theorem, 21
Liebnitz law, 37, 58
Lifshitz, E. M., 11, 34
Limber, D. N., 77

Linear momentum
 conservation, 44
Local angular velocity, 59
Lorentz force, 16, 48
Lorentz frame, (see Lorentz space)
Lorentz metric, (see Lorentz space)
Lorentz space, 35, 55, 57
Louisville Theorem, 9, 20
Magnetic disruption energy, 90,113
Magnetic energy, 69, 113
 variation, 73-76
Magnetostriction, 47
Mass,
 acceleration of center, 32
 conservation, 9,16,31,36,64,98
Matter density, 9,16,41
Maxwell, J. C., 2, 19
Maxwell's laws, 58, 100
Meltzer, D. W., 104
Milne, E. H., 53, 94
Misner, C., 35
Moment of inertia, 12
 relativistic, 36, 106
 variation, 64
 about an axis, 67, 113
Momentum, (see angular and linear)
 conservation, 99
Momentum density, 16
Neutron stars, 103
 stability, 116-118
Newcomb, S., 125
Non-inertial coordinate frames, 48
Ogorodnikov, K. F., 19
Oppenheimer, R. J., 105
Ostriker, J., 94
Parker, E. N., 2, 27
Phase average, 19-22, 86
Phase space, 9, 20
Plancherel, M., 19
Poincare, H. I., 2
Poisson's equation, 10, 42
Polytropes, 34,109,114,119
 Kelvin-Helmholtz contraction
 time for, 122
Post-Newtonian approximation (see EIH)
Potential, 10,29
 Gravitational, 14
 Rotational, 49
 Scalar, 31,43
 Vector, 43
Potential energy, 2,14,25,47,84
 average, 2
 Gravitational, 97,113,123
 variation, 65
Potentials,
 relativistic, 43
 relativistic - interpretation,
 43,45
 scalar, 31
Pulsation,
 effects of surface terms, 78-81
Pulsation energy, 69
 variation, 70
Pulsation frequency, 87
 rotational effects, 76
Pulsation periods, 68
 effects of rotation and
 magnetism, 77

Pulsational stability, 89
Pulsational frequency,
 for determining stability, 87
Rayleigh, Lord, 2, 87
Relativistic matter density, 41
Relativistic moment of inertia, 36, 106
Relativistic correction terms to
 energies, 45, 106
Relativistic equations of motions, 42
Relativistic kinetic energy, 58, 117
Relativistic form of Lagrange's
 identity (see Lagrange's identity)
Relativistic potentials, 43, 45
Roche model, 113
Rosenthal, A. 19
Rotational energy, 52
 variation, 71-73
 kinetic, 69, 72
Rotational potential, 49
Scalar potentials, 31, 43
Schwarzschild radius, 104,109,111,117,
 119,120
Sciama, D., 126
Secular stability, 85, 93
Siniai, Y., 21
Space,
 Lorentz, 35, 55, 57
 phase, 9, 20
 velocity, 9
Space-time, 35, 124
Stability:
 global, 88
 secular, 21, 93
 against magnetic fields, 90-93
 against rotation, 91-93
 of neutron stars, 116-118
 of white dwarfs, 109-111
Stability criterion,
 Jacobi's, 86, 117
 Jeans, 95
 pulsational, 89
Surface terms:
 magnetic, 81
 pressure, 81, 121
Thermal energy, 69
Thorne, K. S., 35, 104
Time averages, 14, 19-22
Time variation, 100
Tooper, R. F., 104, 111, 118
Total energy, 56, 84, 86
 variation, 108, 112
Variation of (see specific term)
Vector potential, 43
Velocity,
 angular, 48
 average stream, 9
 equatorial, 115
 local angular, 59
Velocity space, 9
Virial, 2, 4, 17, 18
Virial equations, 31-33
Virial tensor, 29
Vis viva, 2
Viscosity, 16, 94
Volkoff, G., 105
Volume - average, 111
Wheeler, J. A., 35
White dwarfs, 34, 104, 110
 stability, 109-111

RAYMOND H. FOGLER LIBRARY
DATE DUE

SUBJECT TO
WEEKS